The THEORY *of* MATERIAL MIND

The THEORY *of* MATERIAL MIND

Or the rediscovery of metaphysics

A description of the mechanics of mentation and their role in the origin of paranormal phenomena.

COMPLETE VERSION

Philip Hodgetts

Copyright © 2017 by Philip Hodgetts.

Library of Congress Control Number: 2017900649
ISBN: Hardcover   978-1-5245-9728-3
              Softcover   978-1-5245-9727-6
              eBook   978-1-5245-9726-9

All rights reserved. No part of this book may be reproduced or transmitted in any form or by any means, electronic or mechanical, including photocopying, recording, or by any information storage and retrieval system, without permission in writing from the copyright owner.

Any people depicted in stock imagery provided by Thinkstock are models, and such images are being used for illustrative purposes only.
Certain stock imagery © Thinkstock.

Print information available on the last page.

Rev. date: 10/06/2017

To order additional copies of this book, contact:
Xlibris
800-056-3182
www.Xlibrispublishing.co.uk
Orders@Xlibrispublishing.co.uk
742072

CONTENTS

Introduction xi

Chapter 1: The universe 1
1. The interconnectedness of the universe 1
2. The god within 1
3. The materiality of mind 3
4. Bioplasma (BP), psychoplasma (PP) and the bioplasmatic envelope (BPE).. 3
5. The Kirlians' Absent leaflobe 10
6. Psychoplasmatic recordings 12

Chapter 2: The god within and the primacy of feeling................. 15
1. The god within (GW) and the unconscious mind.......................... 15
2. The nonreplicable nature of psi experiments. 23
3. The 'artlessness of youth'.. 24
4. The spirit in the heart (S) and the conscious mind. 25
5. The dual nature of reality. 27
6. The origin and nature of thought.. 32
7. Hume, metaphysics and the self. 34
8. The collective unconscious (CU) 35
9. The collective conscious (CC). 36
10. The primacy of feeling (POF). 37
11. Illustrations and assertions supporting the concept of POF. 41
12. The mechanism of judgement and belief. 42
13. More examples of POF recognised or ignored. 47
14. The phantom limb. 54

Chapter 3: Godel's theorem and to feel is to know 56
1. Head, heart and gut. 56
2. Memory 58
3. Kurt Gödel and the Incompleteness Theorem. 60

4. Intuition and instinct. 61
5. Expressions describing intuitively perceived truths. 62
6. To Feel is to Know. (TFITK) 64
7. Examples of To Feel is to Know (TFITK) 66

Chapter 4: The universal unconscious and the afterlife 69
1. Telepathy 69
2. Clairevoyance or second sight 71
3. Psychokinesis (PK) 72
4. Unconscious Mind Suppression Phenomena. (UMSP) (poltergeist phenomena). 73
5. Kinetic Energy expended in UMSP. 75
6. The bioplasmatic envelope (BPE) 78
7. The BPE connects all life that exists or ever has existed. 80
8. The Universal Unconscious (UU). 86
9. Levitation. 88
10. Out of the body experiences (OBE) and consciousness. 89
11. Teleportation of an object, living organism or human being. 91
12. Dowsing and divination 94

Chapter 5: Extended youth and the elixir of life 96
1. The anthropic cosmological principle. 96
2. Unqualified atheism is no longer tenable 98
3. The god within and true love invalidate atheism 102
4. Freewill and the housefly 103
5. Mental and emotional activity and psychoplasma 103
6. The dual nature of imagination (DI) 105
7. Extewnded youth (the alchemists' elixir of life). 112
8. Ghosts and their nature.(most tenuous at PP = Basal density – quite invisible) 117
9. The Conscious Mind is Finite. 120
10. The biological field (BF) 121
11. Making a distant person turn round by staring at his back 123
12. The non-distance-related nature of psi 124
13. The Einstein/Podolsky/Rosen paradox.. 126

Chapter 6: Love and psychoplasmatic bonding 129
1. The Sexuality of the God Within 129
2. Falling in love 129
3. Omni-love and self-love 130
4. Time, space and the afterlife.(See addenda 42) 131
5. The universal conscious 145
6. Precognition 145
7. Psychometry or token-object reading. 151

Chapter 7: Physiological functions and group minds 154
1. The function of the sneeze. 154
2. The function of the cough 156
3. Clearing the throat. 157
4. The significance of the burp 157
5. Predictions and observations made by the theory confirmed as true 157
6. Discoveries and elucidations made possible by the theory. 158
7. PP bonds between people, animals and vegetation 161
8. Group minds 162
9. PP blobs, 'streamers,' 'tubes' and 'funnels' 163

Chapter 8: The Enlightenment and Beyond 166
1. Balance, the 'pendulum', and rediscovery of the 'old knowledge'; 166
2. Witchcraft and the occult 168
3. Isaac Newton, deism and The Age of Reason. 170
4. What are we? Where do we come from? Where are we going? 171
5. Evolution was, but is no longer, teleological. 173
6. Paranormal phenomena experienced by the author. ... 174
7. The quantification of metaphysics 195
8. The vacuum wheel psychoplasma detector 197
9. The umbrella PK demonstrator (ref. 9c) 198
10. More on the difficulty in replication of experiments in psi. 199
11. An example of Mutually beneficial feedback between applied science and pure research. 200

Summing Up	203
Diagrams and Illustrations	209
Glossary	211
An Aid to Assimilation of Concepts and Ideas	215
References	225
Addenda	239
Summary Analysis	289
Metaphysics	291
Epilogue	293
Bibliography (Abbreviated)	295
About the Author	297
Index	299

Sit down before fact like a little child, and be prepared to give up every preconceived notion, follow humbly wherever and to whatever abyss Nature leads, or you shall learn nothing.

T. H. Huxley

INTRODUCTION

I am not qualified in any scientific discipline though I did complete four years of a five-year course leading to a degree-level qualification in telecommunications and electronics, in which sphere I spent the first ten years of my working life. My actual work was the development of electronic instrumentation and servo-mechanisms, connected with the aircraft industry.

An unsatisfied longing for adventure brought my brother- in- arms, a journalist, and myself, to the capital as a first stop in the young man's proverbial 'journey round the world'. It was not to be. Love stepped in and he disappeared in the direction of Bognor Regis with a lady on his arm and I found myself alone in a stop-gap job as a TV repair man. This did nothing to satisfy my intellectual and creative longings so, being blessed with a talent for drawing I decided to switch careers and train as an artist, so here I am, a professional, practising fine artist with a host of other interests, specifically and mainly, parapsychology.

The theory contains a lot of material on the afterlife - which I have discovered to exist - and for which I've found an interesting 'quasi-proof', and the rather exotic phenomenon of teleportation, ("beam me down, Scotty"!) which I have found also to be perfectly possible, even straightforward in theory.

Before proceeding further I would like to make it clear that I did not set out with the express purpose of discussing the existence of the afterlife, or the possibility of teleportation. These matters came up simply as a result of my main discovery, that both the conscious mind and the unconscious one are material and extend into the space around the body, connect with other minds, and are composed of two substances which I call bioplasma and psychoplasma. An equally important and actually more fundamental discovery – and here is where I risk being laughed out of court - is that there are two metaphysical entities in the human body, which deploy and motivate the conscious and unconscious minds, namely and respectively, the spirit in the heart and the god within.

It is not generally known by parapsychologists, I think, that the great J.B. Rhine who was the first to put the study of psi on a secure scientific footing, concluded from his researches that there must be a metaphysical component in the constitution of the human being (ref. 10a and 10b) which I quote here:

(a) The question is merely, 'Is there anything extra-physical or spiritual in human personality?' The experimental answer is yes. There now is evidence that such an extra-physical factor exists in man. The soul hypothesis as defined has been established, but only as defined:

(b) What has been found might be called a psychological soul. It is true that, as far as we have gone there is no conflict between this psychological soul and the common theological meaning of the term........The first step was essential, however modest. It has established a point that millennia of argumentation have failed to make. This beginning represents the turning ***of three centuries of domination of our science of human nature by physicalistic theory.*** (Author's italics). It will eventually have the most revolutionary significance, though the full effect may be slow in being seen. The turning of tides is never sudden. End of quotation.

This fact constitutes the essential difference between parapsychology and science in general, which, supported by philosophers such as David Hume, and the advent of the 'age of reason' in the seventeenth and eighteenth centuries, relegated metaphysics to the status of delusion and self deception.

I happened upon a great part of the raw material, the empirical data of my theory, as the results of a psychological trauma which led to a period of acute loneliness, brought on by my adoption of a hopelessly idealistic philosophy of living. During this period various strange things began to happen, and being already interested in such things I tried to find in them some explanatory and, hopefully, unifying structure.

The raw material unearthed by my crazy, lonely life, although painful, put me in a distinctly advantageous position vis-à-vis other researchers. They were researching the experiences of others whereas I was dealing with my own private and extraordinarily revealing ones. Having discovered the main elements of my theory, I felt like a dog whizzing around with a juicy bone between his teeth, of which he simply **would not** let go. I knew that I had the truth in my hands and that I was on the right track. I saw, however, that I would have to relate my undoubted findings, my knowledge, to other phenomena as I encountered them in the course of my daily life, to build up a theory. This I have done, and here is the result to date. I still find myself making more and more mental connections, seeing more and wanting to add to and refine the theory, but I can see this going on for many years yet, so I am presenting my theory as it stands, knowing that it is only the basis for a far more comprehensive and sophisticated treatment by myself and others in the years to come.

Now we come to the difficult bit. My theory, which is the result of fifty years of thought, reading and a lot of personal experience, requires the acceptance of a number of concepts, already touched upon, e.g. the discoveries of J.B. Rhine which assert the existence and reality of the soul, plus my own contribution to the array of metaphysical entities which inform the body and Nature in general, namely, The God Within. This may appear somewhat unscientific insofar as it claims and asserts the reality of intelligences and divinities which abound in Nature and in ourselves. The nature of these entities is apparent from their behaviour which always tends to promote harmony and good feeling. It was suggested by friends that I should refer to them as 'creative forces' but I couldn't since this would be dodging the fact that they really are intelligent and divine, that is, they are divinities, and possess all the attributes thereof, and as such are truly good. Also, as gods, their activities transcend the boundaries of space and time, which are in fact themselves questionable concepts . They are naturally warm, expansive and gregarious, tending to join forces, which means that taken as a whole they are omniscient but, I'm afraid, not omnipotent, since each god is obliged as part of its function, to obey the dictates of the particular heart or spirit to which it is bound, which is not always above directing it into immoral actions, self- aggrandisement and worse. That is, each god has access to the total knowledge and power of the group, whose numbers run into billions, the population of the planet, one to each human being. Each higher animal also has a god within, which behaves in it's own unfathomable and mysterious way, and, I believe with William Blake, that holiness, (**not** a creator God but something which is to be approached with reverence and awe), in some form is found in all forms of life. This also means that learning to recognise and analyse their behaviour is fairly straightforward but their origin and inner nature remain essentially unknowable, a mystery, which necessarily places limits on the extent of knowledge to which science can aspire. This means that the 'grand unified theory' (GUT) sought by physicists is impossible since even if a theory were to be developed which unified the three fundamental physical forces, and gravity, it would still be neglecting parapsychological phenomena, which are all real and actual, though not always capable of demonstration, or experimental replication, the reason for which I have discovered and included in the essay. I do believe, actually, that physics is basically on the right road to the truth, insofar as it is beginning to accept the reality of metaphysics, but it would be the same truth that I, through a series of chance occurrences, hard times and a wide open mind, am now promulgating.

The Theory of Material Mind, taken as a whole, is largely an analysis and description of the 'nuts and bolts', of the physical and metaphysical aspects of the human organism and the way in which it relates to others and its environment, and is as close, I think, as we shall ever get to the 'grand unified theory' sought by physics. I should make it clear that the subject is looked at from a metaphysical, spiritual and conceptual point of view rather than a holistic one. This may give the impression that the theory is in parts a little naïve, even perfunctory, but this is not so. It means simply that due to my limited knowledge of human physiology, I have not been able to explain exactly how the god within and the spirit connect or interface with the body. In a completely holistic treatment the metaphysical and physiological aspects of the human being would be related and explained as a mutually dependent and integrated whole.

This is a task for the future and it will probably take many decades, even centuries, to complete, assuming that is, that it will be possible, i.e. allowed by the god within. Finally, I would like to anticipate and disarm the scepticism which my theory is bound to encounter by giving a quotation attributed to Albert Einstein. "If an idea is not at first absurd it has no hope.

Philip Hodgetts. London 2017.

CHAPTER ONE

The universe

1. The interconnectedness of the universe; 2. The god within; 3. The materiality of mind; 4. Bioplasma; psychoplasma and the bioplasmatic envelope; 5. The Kirlians' absent leaf lobe; 6. Psychoplasmatic recordings;

❖ 1. The interconnectedness of the universe
It is to be taken as an axiom, not a hypothetical or heuristic one, but a real one, that the universe is a space-wise and time-wise interconnected whole. Everything hangs together. All things, physical, mental and spiritual interact. Everything in the past, present or future, is connected, and more or less directly affects, every other thing. Here is linked with there and now is linked with then.

Each private universe or world, overlaps with that of the others to form common ground which is the 'public' universe, real, and familiar to all of us, and each private universe contains phenomena of which inhabitants of other private universes are unaware. All this is surrounded and contained by the Universal Unconscious of which no-one is aware. These things will be described later in the essay. (diag.1).

❖ 2. The god within
There are, in the human body, not only physical or material organs, heart, lungs, glands, intestines and such, but two additional entities which are entirely metaphysical. They have no physical existence whatsoever. They are the god within and the spirit. The first is based in the abdomen and centred on the solar plexus while the second is centred in the heart and expands throughout the rib-cage into the neck and head. (fig.2)

The existence of the spirit, an essential part of religion, (which I see as an obstructive anachronism), was brought into doubt by the rise of science starting with Roger Bacon, Galileo, Newton, Leibniz, Hooke, and philosophers such as David Hume, aided and abetted by Darwin and the Victorian scientists, Michael faraday, James Maxwell, Heinrich Herz and so on. (Ref.10a.b.)

The gods within and, also, possibly, the gods without if such entities exist, and their ramifications, are the unvarying absolute of all life on the planet, past or present. One's own personal metaphysical absolute, and that of all other humans and higher animals, is the god within (GW) which is located in the gut, and which is a vital, active presence whose chief characteristic, which includes its innate and unfailing goodness, is enthusiasm. It is the engine or motor of the unconscious just as the spirit is the motor of the conscious mind. It never tells a lie, doesn't make mistakes and is the active element in the so-called 'gut-reaction'. It is also the locus of sexual and other appetites and in matters of love it is incorrigibly promiscuous and omni-sexual, but in this respect, as in all others, it must follow the dictates of the conscious mind (spirit in the heart plus head), so its sexuality will obey a personal or moral directive, that is, the person will be either hetero, homo or bisexual (or other). Expressed simply; any given person or higher animal possesses:

(a) a spirit in the heart which works in tandem with the head to form the conscious mind, and:
(b) a god within which resides in the gut, and of who's presence, its owner the person, is normally unconscious.

The god within works in tandem with the spirit in the heart and the head i.e. the conscious mind, and obeys its every instruction. That is to say, for instance, that when I go to make a cup of tea (in obedience to a bodily desire), the physical energy will be supplied by food in the stomach and gut reacting with acid and enzymes, but the emotional or metaphysical energy, the ***desire for*** and the ***feeling*** of relief or elation at the thought of a rest and a cup of tea, will be supplied by the enthusiastic god within. The god within is infinite, transcends the laws of physics and operates beyond space and time. It is responsible for the faculty of intuition which is the matrix of all creativity.

❖ 3. The materiality of mind

The mind as conceived of hitherto, has generally been imagined in all kinds of shapes and guises, and my conception of it is as an entity possessing a conscious component and an unconscious one, as does the presently accepted orthodox one. My concept, however, is of a material mind whose both components are composed of a substance; the conscious one I call 'psychoplasma' (PP) and the unconscious one 'bioplasma' (BP). Each mind is animated by a metaphysical agency, the conscious one is the spirit which is the basis of personality, and the unconscious one is the god within (GW) which is the locus of desire and enthusiasm. In such a mind it will be seen that spatial concepts, e.g. tables and chairs etc., and thoughts and feelings are larger or smaller volumetric forms in psychoplasma while other things such as ideas or perfect geometric figures, are immaterial, i.e. metaphysical. This latter includes of course the two metaphysical entities the spirit(S) and the god within (GW).

This model of the mind, was discovered, not invented, just as was the circulation of the blood by Harvey and the unconscious mind itself by Freud. It is not a hypothesis, but a real, existing and substantial (material) thing which has grown in tandem with the body, through all stages of evolution. The fact of its being largely material and closely integrated with the body leads on to some extremely strange and unexpected revelations. (See addenda items 47 and 48).

❖ 4. Bioplasma (BP), psychoplasma (PP) and the bioplasmatic envelope (BPE).

In the past philosophers sometimes referred to the concept of 'mind-stuff' in the absence of a clear idea of what and where the mind was. I myself spent many years pondering and making observations on this question and one evening in 2004 I think, (recorded in my journal but hard to locate) I was rewarded with a dramatic revelation, averitable epiphany. I was thinking about the spirit (on which more later), as a presence in my heart, having already reached the conclusion that all mental activity originates there, when I felt a sudden uprush from this locality, the centre of my chest, and it was as if a gushing stream of clear, airy, semi-gaseous fluid flew up and encapsulated my head and upper body. I realised immediately that this was the stuff of the mind, upon which I had been brooding and pondering so long. I realised later that it was only part of the mind, the spiritual basis of it, and that there was an additional, denser, heavier fluid, created, I believe, in the cells of the body.

This stuff of the spirit (S), I realised, was the basis of consciousness, and the cell-generated stuff, which I call bioplasma (BP), was the stuff of the unconscious mind. The spirit never exists on its own, when existent in a material body, but is always combined in a mixture or compound with BP to form a substance which I call psychoplasma (PP). This mixing takes place mainly in the body, I believe, and is exuded through the skin as PP. The spirit on its own, I think, is incapable of affecting inanimate matter but operating as PP it can do exactlythat by virtue of the materiality of BP which is one of its constituents, and when it does the phenomenon is called psychokinesis (PK), the conscious, voluntary exercise of 'mind over matter'. When I say the 'conscious, voluntary' exercise I don't mean staring at the object concentrating hard on 'willing' it to move, it is more, for me at any rate, a matter of relaxing and 'wanting' or waiting for the reaction. I quickly realised that BP, generated in the cells of thebody, must be generated in all living cells and connects together every speck of living substance on the planet (a fact which is deducible, in hindsight, from the information related to Ref.6) thus forming an Earth-enveloping blob which I called the Bioplasmatic Envelope (BPE) (ref.6). This, mixed or compounded with (S) to make psychoplasma (PP) is the fluid posited by Arthur Koestler by means of which telepathic communication and much, if not all, other psi takes place. (ref.4a). It is also the material embodiment of Carl Gustav Jung's 'collective unconscious'. The sum total of BP + S = PP on the planet, I call the psychosphere' (PS). Although the unconscious works through BP, in its telepathic and other psi activities, and the conscious mind works through a compound of BP plus S (i.e. PP), seen objectively both the unconscious and the conscious minds operate through the medium of psychoplasma (PP) since BP and S will have mixed to become PP which changes its shape, density and BP to S ratio according to the circumstances, to saturate and totally surround the body especially the head and thorax

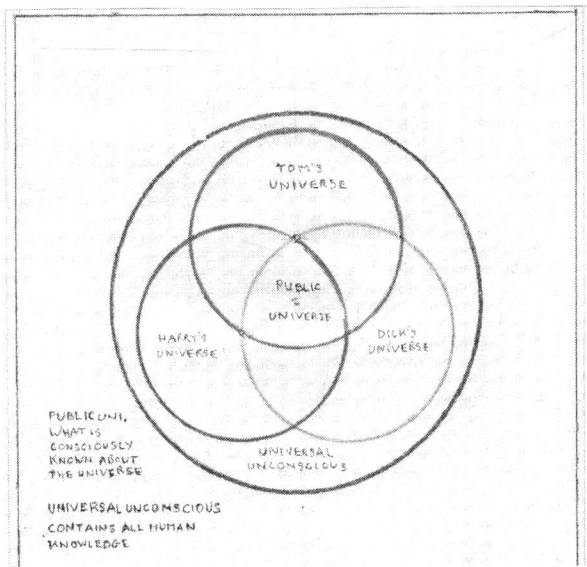

Diag. 1: The universe

NOTE: The ratio of BP to S will decrease with altitude (since BP is material, subject to gravity, and much denser than S), so that at, let us assume, eight kilometres above sea-level the density of the CC will be at the basal level (see chapter 5 item 12). Any further increase in altitude and it will drop sharply to zero, rather than fading away, thus defining the boundary of the BPE, which can be visualised as a large, shimmering soap-bubble containing, not only air, but PP, enveloping the earth.

The BPE, though permeating and connecting all living things, things by which it is generated, remains essentially the medium of the unconscious. The terms BP plus S in the little equation just now referred to, add to make PP, the medium of the unconscious mind (BP) plus the conscious one (S) and which also, of course, permeates all living things, and unlike the BPE it also permeates the greater part I believe, if not all, non-living things, rocks, water, metals, crystals above and below ground, man-made artifacts, dice, furniture, clothing, rooms and public spaces, glass, plastic etc. They are all embraced, filled, or permeated by the spiritual component of PP depending on circumstances, and this fact I think, is central to all parapsychological phenomena. The only part of the earth that will not be so permeated, I suspect, is the earth itself below a certain depth although this could be several thousand metres. This point will be further explained in Chapter 4 (item 12) and Addenda (item 26)

It seems probable, as explained just now in different terms, that the basic, pure BPE, although still existing in its entirety, will be suffused throughout with S though the ratio of the quantity of BP to that of S, (BP:S), will vary depending on terrain and other circumstances. I think, as explained, that the outer surface or topmost level of the BPE must remain as I perceived it in my first astonishing vision of it - i.e. a sharp cut-off point rather than a gradual fading away, and until this point of cessation is reached, the density of the BPE plus its admixture of spirit must vary between maximum possible and basal as in graph 1. I think, also, that the ratio of S to BP (S : BP) will be proportional to the altitude at which they are measured, but this quantity too will depend on the propinquity or otherwise of other living organic material.

I think that PP, as the medium of the Collective Conscious which can be loosely defined as the sum of all conscious minds (more exactly defined in chapter 2), must extend to a height of seven or eight kilometres. It is clear that birds must be enveloped in it, and even passengers in jet aircraft. It is a fact that there are people living in the international space station, which is so far from the earth's surface (220 miles) that one would think of it as being ouside the planet's PP, but I think in fact that it must remain connected by a long 'funnel' of PP (see chapter 7 item 9). It would, in any case, be interesting to do a series of telepathy experiments between the ISS and Earth: if they proved negative, giving only chance results, this would suggest the idea that in these circumstances, there would be a small 'mini-BPE' which connected the astronauts aboard the ISS. This also could be checked by doing telepathy experiments between crew members which should prove positive, that is, better than chance. One should bear in mind, however, whenever proposing an experiment in the paranormal, that such experimentation is always open to the mischievous refusal by the god within to oblige and by doing so to withold its power, a power that is the very essence of life itself. (Chap.8 item 10, and Chapter 2 item 2). I used to believe, and still do I think, that an organism which was not included in the BPE, or at least, a 'mini-BPE' could not survive, which raises the question of Yuri Gagarin's solitary orbit of the earth; it is clear he must have stayed attached in some way, via the spiritual component of BP, probably a 'funnel' or 'tube' (see chap. 7 item 9).

While we are on the subject of funnels and tubes I might as well reveal another recent insight, into 'astral travelling', a practice embraced by certain students of the occult, and which I think is simply an extension of the OBE. It has occurred to me that if the subject/spirit in an OBE floats or travels more than 50 to 100 metres away from the body, the PP which keeps it connected with it must, or may, assume the form of a long, shimmering cord exactly as described in literature on the occult, and the cord's cross sectional area would be inversely proportional to the distance involved, I would think . The cord's surface would be at the basal level which is why it would shimmer. It is obvious, as expressed in the literature of the occult, that if this cord is broken the organism dies and the spirit goes to the afterlife followed, I would think, by its permanently attendant god within.

Bioplasma, as stated, and psychoplasma too, is a clear fluid with a consistency which varies between thin, watery/sticky on the skin via a light, clear, consistency to very fine and rarefied. Its consistency also varies in accordance with the identities of the people involved, and the existing circumstances; it mixes with the substance, if such the spirit (S) be,(see addenda 109) emanating from the heart, to make psychoplasma as already written .This, PP, is the stuff of the conscious mind and it is material - hence the title of the essay - a substance possessing mass, probably only because of its proportion of BP which is definitely material. Like BP its nature is myriad and it can exist in many forms, and although its usual state varies between thin and sticky near the body, and tenuous, its density can vary between these values to the basal density, which is only a fraction of that of the air in which it floats. During the last stages of a teleportation (which I am sure is possible and shall describe later) it would gradually (or instantly see chap. 5 item 13) 'condense' to become the materialised flesh, blood and bone of the teleported body. It can sometimes be felt on the skin as being cool and slightly sticky and it can vary, depending on circumstances, between the forms of a tenuous placid lagoon or rolling ocean. When one wakes from sleep in the morning and slowly comes to consciousness, the sleepy, thick and groggy state that one may find oneself in, I think, is simply the mass of PP that the cells of one's body and the spirit have generated during sleep, in fact it could be that replacing psychoplasma is one of the important functions of sleep. At first the reader may find it difficult to perceive, and thus believe in the existence, of PP, but if he bears in mind the logic of what I have written and is patient, he may find himself becoming more sensitive to its presence. The first revelation of the presence of the BPE came to me five or six years ago, but it was only last year, in 2010, that I clearly observed a group mind – a beautiful experience. (Chapter 7 item 8).

Sometimes the mind, at times of stress and insecurity I think, loses the vague softness of its boundary where it fades into the CC and its boundary to become sharp and clear-cut like the surface of a soap bubble, (the surface of the BPE also has this quality) and in so doing must, I think, be cutting itself off from the possible dangers inherent in free association with its neighbouring minds, although, I think, it must always retain a tenuous connection with one or more of them. (fig.7). PP I would think, will be male or female depending on the gender of the individual in question. This requires more thought i.e. the two defining sexual poles are male and female but there would be, I think, a whole spectrum of of relative strengths between these opposites so that the PP of the Collective conscious would be neutral, as would be, without doubt, that of the Collective Unconscious. Although it is material, PP is usually tenuous although it can exist at any density up to thin and sticky in and near the organism. All substances,. as far as I know, brick, wood, plastic, glass, lead are transparent to its spiritual component.it in its more tenuous states. If one wants to weigh air one extracts all the air from a container to create a vacuum which will tend to make it rise and then notes how much weight must be applied to hold it down. One cannot do this with PP because it would permeate the walls of the container. I have, however, devised an experiment which may make it possible to demonstrate its existence. It involves the construction of an instrument which I call 'the Vacuum Wheel PP detector', which should make it possible to detect the presence of BP and PP. Details of its construction and mode of operation appear later in the essay. This device was conceived long before

I discovered the reason for the difficulty in replicating psi experiments, and may, therefore, be subject to the same limits.

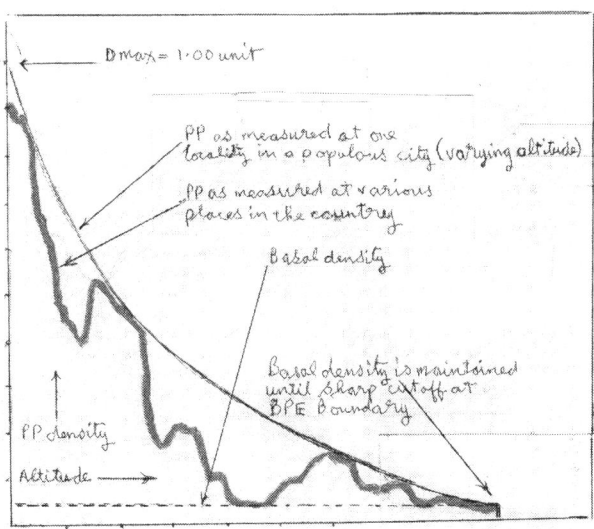

Graph 1: Effects of altitude and locality on psychoplasma density.

My mind, therefore, originates in my body which it completely surrounds, and extends into the surrounding space mixing with other minds as it does so and/or merging into the collective conscious. It may be helpful, though perhaps unnecessary, to inform the reader that as I describe the phenomena of BP and PP, he/she should be understanding that he is himself immersed in a sea of PP just as a fish is immersed by the sea in which it swims except that the PP permeates the body while the sea does not permeate the fish. He should, of course, also understand that this plenitude of PP saturates his/her body, and permeates the walls, floor and ceiling of the room wherein he sits, to join with and become part of the CC and the CU. And lastly he should understand that he is connected to all the life in his immediate locality. The people next door and those upstairs, the mice under the floorboards, if there are any, spiders, flies, woodlice in the woodwork, squirrels, robins, blackbirds, in the trees outside, earthworms in the garden, and micro-organisms in the kitchen sink outlet pipe; all of these organisms are connected via PP and are mutually affecting to a greater or lesser extent.(ref. 6).

Here is an account of an experience which should help to understand the nature and function of PP. Many years ago, I was feeling very low in spirits and could see little point in being alive. I was driving towards Welshpool near the border of Wales. Suddenly I saw a beautiful girl standing on a hillock some yards away from the roadside, and my spirits lifted. I saw beauty, friendship, love, meaning – life was not bottomless and insubstantial but, as revealed by the girl, was full of meaning and love, enveloping all things,.. substance …bioplasma, and, in tandem with spirit……psychoplasma! (It doesn't sound very romantic but BP is the very stuff of love and desire and it underpins our every action). Psychoplasma, which is inescapably involved with sexuality, is about living and creating life and is fundamental and essential throughout the living world. Psychoplasma is responsible for the warm feeling of friendship that one gets from one's social contacts and its absence is the cold feeling of loneliness.

❖ 5. The Kirlians' Absent leaflobe

My first introduction to the phenomenon of bio-plasma was through the work of the famous Semyon Davidovitch Kirlian and his wife Valentina. They developed a system of photography using, instead of light, a high-voltage alternating electrostatic field. This system has been used to produce colourful pictures of the 'aura' of plants and other living things. The most dramatic, as far as I know, of these experiments, was an occasion when they irradiated a leaf which had five clearly defined 'lobes'. After the irradiation they cut off one of the lobes, and were surprised to find that in the space previously occupied by it, there was a faintly glimmering, luminous image of it. This suggested that the plant had some kind of spirit or 'aetheric body' to use somewhat outdated language (fig.1).

They gave the name 'bioplasm' to the substance of which the 'spirit body' was made, and suggested also that it could be a fourth state of matter. It seemed obvious at the time to assume that it was identical with the substance that I had discovered which I therefore also called bioplasm (BP); but since 'plasm' denoted a solid substance I felt that 'plasma' was more accurate for my purposes, since BP is a fluid, or, depending on circumstances, something between the two. Its density varies. It is generated, I believe, at cell level in the human body, and, in fact, by all cells in all living bodies. As with the air above and around us, we are all surrounded by it, though it is congruent mainly with the upper body, and to a certain extent, the lower body (although this varies with the circumstances). It originates in our bodies, and all cellular life, as a mixture or compound of BP and spirit where it forms a mass whose density varies with the current activity. It is exuded through the skin to pass through walls, and under the ground to a depth of possibly thousands of metres (collective and local unconscious) and suffuses living matter forming volumes, and odourless masses big and small - small, dense masses inside larger, softer masses and all enveloped by the collective conscious. It is thus that clear, sticky masses with the tenuity and delicately penetrating, but odourless quality of the emmissions of camphor oil hover and float around individuals, ebb and flow back and forth reaching seven or eight kilometres into the air above us to form, oddly enough, a sharply defined PP boundary. The density at this boundary is the basal density, and the reason for its presence is given in chapter 5, item 12.

Fig 1: The Kirlians' missing leaflobe

PP can sometimes be felt on the hands and the face, as already described, as a light, cool, 'sticky stuff'. I owe this expression to Kate Bush who mentions it in one of her songs. It is much more plentiful in young people than older and it is strongly related to sexuality. One's physical state of being is strongly related to the amount of PP contained in one's body at any given time; if one is full of PP one is full of feeling and substantiality. If one is short of PP one will feel thin and insubstantial. I think a boxer 'shaping up', getting ready for a session with the punch-bag, is amassing PP in his arms and shoulders to give his punch more momentum. This suggests an experiment that could be performed on such a person i.e. (a) measure the impact pressure of his punch before and after 'shaping up'. (b) compare them. If (b) is greater than (a) this would support my judgement of the function of his actions.

The late Kenny Everett, broadcaster, used to strive to utter the most nonsensical but amusing ideas and images that he could summon. e.g. 'the lady had thighs like tugboats'. I empathise with him because I myself used to do this, in mock verbal contests with friends, 'insulting matches' for instance, in which one hurls insults at one's opponent. The subjective experience of doing this was to reach up and behind one's head into one's 'mental field' with an imaginary arm and hand, to grasp a handful of nonsense. (A handful of PP in fact. I now see this clearly).

❖ 6. Psychoplasmatic recordings

Many people have reported seeing centuries-old streets and buildings one day only to find that next day they were gone. These occurrences are called 'time-slips'. People also report having 'unpleasant sensations' in certain spots as if something 'bad' had happened there and left traces behind. It has been suggested that brick walls and even the ground itself record events at sites of extreme emotional experiences such as murders, suicides and battles. I would like to suggest that what actually happens is that very dense, or even solidified, PP, detached from its related mind(s) is lodged in the walls and the ground, and when contacted by a modern mind becomes temporarily, or permanently, part of it, thus yielding up sensations and percepts of the long-deceased protagonists in the recorded events. I do not as yet know the precise ways or mechanisms of how PP operates, it seems obvious that it is full of information in some way so that when a portion gets detached from the main mass of the subject's mind to become lodged in a wall or the ground at some isolated spot, or, more likely, soaks into the walls and ground to become solid, it may be full of information about the doings and perceptions of its owner at the time that it got detached. It is already established that PP permeates and passes through the walls of a building and penetrates many metres into the ground so this is a very cogent explanation. (It should be noted that I've never experienced this phenomenon myself).

More evidence for this assumption is that I myself, while developing the theory and becoming aware of my 'new mind' and its characteristics, often felt in the turmoil of my difficult days, that pieces of it were being dislodged and lost. Going further it is obvious that the missing pieces, like all the other missing pieces from other injured minds, must have become part of the CC thus making it a rich source of intermixed information to which all and any minds would have access. Even more cogency is lent by an experience I had three or four years ago. I was in my room in the care home where I used to live, trapped in a mass of telepathic attachments to other people in the building whom for various reasons I had to avoid. The PP was not only extremely dense, but felt as if it had solidified around my head and I found myself unable to move for a minute or so. This brings to mind the familiar situation where a person may suddenly feel unwell and need to open a window to get some fresh air. This happens to me very often and I am beginning to believe that what is really happening, is that the person concerned will suddenly become morbidly aware of the overpowering density of PP in the room, and will want to go outside where it is more rarefied. This would probably happen only where there are many people in the vicinity, in the same room or adjacent ones, a situation where there are several people joined by PP attachments. It might be appropriate to note here, in passing, that the theory of material mind conceives of a human being as a duality as did Descartes, that is:

(a) body
(b) mind (spirit +god within.)

(see fig.2). The spirit in the heart combines with bioplasma managed by the god within to produce psychoplasma. These two metaphysical entities are shown in the diagram as having clear boundaries but in reality they would extend as BP plus Spirit, that is PP at $d = 1$, throughout the body, to be exuded through the skin to form the material mind. I do not know whether PP is continuous or particulate although I am aware of the work of two Russian physicists, V.S.Grischenko and Viktor Inyushin who, in 1967, reported giving the names 'bioplasma' and 'bioplasmic body' to substances discovered in their research. They reported bioplasma as being composed of ions, free electrons and free protons - in other words - subatomic particles that exist independently of a nucleus. (ref. 7).

CHAPTER 2

The god within and the primacy of feeling

1. The god within and the unconscious mind; 2. The nonreplicable nature of psi experiments; 3. The artlessness of youth; 4. The spirit in the heart and the conscious mind; 5. The dual nature of reality; 6. The origin and nature of thought; 7. Hume, metaphysics, and the self; 8. The collective unconscious; 9. The collective conscious; 10. The primacy of feeling; 11. Illustrations and assertions supporting the concept of the primacy of feeling; 12.. The mechanism of judgement and belief 13 More examples of the primacy of feeling recognised or ignored; 14. The phantom limb.

❖ 1. The god within (GW) and the unconscious mind

Inside each human being and higher animal, as written earlier, and by logical extension, in some form, every other animal, there resides an intelligent entity, the god within, (to which I was introduced by Arthur Koestler who, as stated, also posited the existence of a fluid medium through which telepathy worked [Ref. 4a]).The GW is the origin and source of enthusiasm, (from the ancient Greek; 'enthousiasmos', from 'enthous', 'possessed by a god'). It is located in the gut in humans, centred at the solar plexus, is the epitome of goodness and is incapable of initiating an immoral or evil act of its own volition, but since its actions are subject to, and guided by, the spirit in the heart and the head (which make up the conscious mind), which may not always be as morally responsible as it could be, can become involved in immoral or even evil actions. It operates outside the laws of physics, transcending space and time; whether it needs sleep as does the conscious mind, is debatable. I suspect not – the idea of a divinity requiring sleep is slightly incongruous, and it is a fact that dreams do occur – Freud's royal road to the unconscious – which appear to engage the spirit or 'essential self' and the unconscious in joint activity. It is also a wise dictum to 'sleep on important decisions' which itself suggests that the unconscious, motivated by the GW, does remain active during sleep.If help is needed to convince the reader of the god's existence, one need only observe that the unconscious itself, which everyone accepts, and its activities, happen on an entirely unconscious but still non random level; i.e. there *is* something which purposively motivates the unconscious – that is, the god within.

The god within is the source and basis of friendly and loving attachments between men, women, children, animals, and vegetation (pets, house- plants and gardens), whether of a sexual nature or not, and sexual love and desire. These attachments can sometimes extend even to inanimate matter - material objects -where PP has permeated the object and become slightly more dense. In this respect I believe that it is also the source of love for personal possessions, activities, one's home and job. It is also the engine of creativity and humour in the unconscious mind, of which it is the fundamental, active force or motor, and is responsible for the amassing and deployment of bioplasma, the mysterious substance already described which joins all life on Earth.

The nature of the god within is to join, through the subtle medium of BP, with other gods in other people and, I am sure, animals and plants. The GW, as indicated a few lines ago, is effectively the motivating force of the unconscious, just as the spirit is the motivating force of the conscious mind, in fact I think that it can be almost exactly equated, function-wise, with that entity, except that the unconscious contains additional material in the form of memories stored in various parts, especially lower ones, of the body, although, qualifying this, there is a very large store of memories, of information, places, friends, things learned by rote, (by heart) in the heart or chest the 'locus operandi' of the spirit. The god within is also the agent of humour, one expression of which is the belly-laugh which comes straight from the god itself. Ultimately, I think, the unconscious is not simply much larger than the conscious mind, but is infinite.

In PP, (BP + S), the god within and the spirit join forces to make the conscious mind. This means that the conscious mind could not exist without the unconscious one. In the unconscious mind the god within works through BP alone.

Jean-Paul Sartre, strangely, did not believe in the unconscious but operating without an unconscious would be like trying to drive a car which had no engine. The existence of an independent personal unconscious was confirmed for me by an amusing incident which happened in the bookshop of my 'alma mater'. I was sitting in an armchair relaxing after an extremely exhausting day out of college. Suddenly the word 'snekot' came into my mind and I sat for a minute wondering what it meant. As I sat there my eye lit upon two words painted on the glass window of the door. The words were 'book tokens' and to my amusement and surprise, I realised that my unconscious mind had reversed the word tokens and this had to be because at the time I had fallen into the habit of amusing myself by mentally reversing words to see what happened.

I first became aware of the god within (in myself), during a period of extreme loneliness and incipient mental illness, recovery from which left me immeasurably stronger, sharp and aware. I was aware, during my illness, of 'a voice crying in the wilderness' and it was saying, "all gods are good but some serve evil ends", in other words, there exists, as written a few lines ago, in each one of us a perfectly good god which is incapable of initiating or committing evil or immoral actions, but which may be directed by evil in our hearts, or simply misunderstandings leading to mistaken actions, into supplying physical energy (gut) and metaphysical energy (god within) for the perpetration of immoral, evil or mistaken actions. I have since seen that these thoughts (all gods are good etc.), whose provenance I could not at first locate, but which I trusted, probably came not from the spirit, (the essential self), by which they were perceived, but from the omniscient god(s) within in the collective unconscious as do the words of a newly created song. One might assume that since the spirit is also a metaphysical thing, it too must be a source of energy, and, considering its role in conscious telepathy and other psi phenomena, I think this is true, although the energy it wields, I think, is much less than that of the GW. In the absence of any definition of this other sort of metaphysical energy, conscious energy – the energy of the spirit – one may call it spiritual or psychic energy.

In view of the fact that the god within is associated with sexual appetites, lust and desire, it is interesting to speculate as to how its nature and behaviour would be affected by castration, resulting in the status of 'eunuch'. What would happen to its usually generous and enthusiastic nature?

It is perfectly clear that the unfortunate castrated subject would have no sexual drive or appetite, but a little research has revealed that the status of eunuch doesn't automatically result in an effete, weak, and overly docile individual. It would appear that eunuchs who were castrated post-puberty can display great courage and determination, and, one assumes, enthusiasm, (the GW's chief characteristic which could hardly abdicate) though of a more controlled and responsible kind than that of the non-eunuch as shown in the following quotation from 'Wikipedia':'The idea of the eunuch effectively guarding the household may seem laughable to some. But although eunuchs have endured more than their fair share of scorn and ridicule over the centuries, many ancient cultures acknowledged that despite their literal "lack of balls", eunuchs were not (necessarily) cowards. In fact, the eunuch could often be relied upon to be a fierce and loyal guardian of the master's home and family. Like the masters castrated ox, a castrated man could also be strong and well-muscled - and just as steadfast in the face of danger or adversity. And while a lessening testosterone (through loss of testicles) did not significantly lessen the eunuch's courage or his strength, it did however tend to make him calmer and more even-tempered. (Again, like the ox.) As a manservant he was therefore less prone to outbursts of irritability and rage. But while some castrated men were muscular warriors, the once common stereotype of a eunuch -- the she-male with soft features, a limp dick, a soft body, and a high falsetto voice -- does have some validity. That is, if a male is castrated before puberty, usually before age 10, his body will never experience the normal changes ushered in by the pubescent rush of testosterone As a result the castrated boy will retain many of his childlike qualities into adulthood, including the inability to get a full erection, and a lack of body or facial hair..

My acquaintance with the god within goes back to the early seventies when I learned of its existence or the concept thereof, as written, from the writings of Arthur Koestler. At that time I was teaching at Goldsmith's College School of Art which is part of London University, although at the time I had not yet embarked on a definite program of research – that came later. It was soon after this that I began to feel the effects of the incipient mental illness already mentioned. It took the form of a compulsion to be as ideal a human being as possible and was a rejection of my former 'easily fobbed off' or 'loser' self and one of its effects was a that I gradually became aware that the god within, *my* god within, was real and existent in my gut, as were those of other people in theirs. I also became aware that it supplied the emotional, or metaphysical energy which accompanied, and fuelled, my actions. In view of all this, I classify the energy supplied by the GW as metaphysical since the god within is still present and it is the locus and origin of unqualified love, that is, sexual love, love of friends, family, hobbies, home, possessions etc., that is *all* love except for the conjugal variety. Thus what was considered by me to be sexual energy, can equally be thought of as 'divine energy' or, as just now concluded, metaphysical energy. The spirit too, in its status as the person, him or her self, is a metaphysical thing and although, I think, it appplies or wields a certain, smaller amount of energy in its activities its main attributes are courage, drive, initiative and, in tandem with the head and gut, judgement – all of which are aspects of freewill and the person. Conjugal love is felt in the heart as well as in the gut.

Almost all metaphysical energy comes from the GW, and it is the source of a unifying love of all things, friends, relatives, pets, possessions, activities and is the origin of appetites, and a lust for life while remaining friendly, generous and expansive.. The collective conscious (CC) is much more tenuous than the bioplasmatic envelope (BPE)..The reason for this is the marked difference in the natures of the god within and the spirit. The GW is constant i.e. ***always*** warm, friendly, generous, humourous, enthusiastic and positive, which leads to strong, solid BP bonds and a powerful BPE.(It is also a fact, of course, that fear can be felt in the gut). The spirit, however, in collusion with the head as the essence of the subject's character, can often be cool, distant, unfriendly, scared, generally negative and subject to moods, which means that the bonds between conscious minds will fluctuate in strength and the CC, which is an amalgam of conscious minds plus a 'no man's land' of generally accessible PP, will be much more tenuous and less powerful than the unfathomed and all-pervading BPE.

This difference between the collective conscious and the bioplasmatic envelope is responsible also, for the fact that the proportion of bioplasma to spirit varies. The BPE will be located, in the main, at ground level to hundreds or thousands of metres below it, and will have a much greater proportion of BP to S than will be found in the CC because of the strength and density of the bonds between the gods within, and the fact that BP, being a physical substance, will be held near to the earth by gravity, while S, its spiritual component will probably remain more or less constant whatever the altitude at which D is measured.(assuming that one can or could measure it). The BPE is much denser than the CC but does not represent the totality of psi phenomena, since the spirit itself is an essential component of the psychosphere (PS) which contains all psi phenomena on Earth including the CC and the BPE. This means that the ubiquitous BPE will occupy the same volume as the PS but will be suffused with a proportion of spirit. The surface of the PS, as seen in my first 'vision' of the BPE, is shiny and discontinuous as is the shimmering surface of a soap bubble. This is because there is a certain density of psychoplasma, the basal density, below which PP does not exist. As written elsewhere, I have arbitrarily given this as $D_{basal} = 10^{-3}$, one thousandth of the full density of 1.00 as it is in the body. This results in a 'cut-off point' (basal density), on the graph of density v. altitude (graph 1). As already described BP originates, I believe, in the cells of the body and it is amassed by the god within to form the personal unconscious (PU), a vast reservoir of memories and experiences which are sorted, arranged and deployed by the god within. An aspect of the nature of the PU directed by the god within is to expand and actively seek out and join, via the medium of BP to other unconscious minds to form the material embodiment of C. G. Jung's collective unconscious (fig.3). The god within which motivates the unconscious, is alive and knowingly, purposively, goes about its business of which, of course, in normal circumstances the subject is unaware. All gods within, I believe, share the same basic, androgynous character. It is warm, enthusiastic, friendly and its nature is to join with others, which explains its manifestation in social situations as 'infectious, humorous, mischievous, expansive and

The Theory of Material Mind

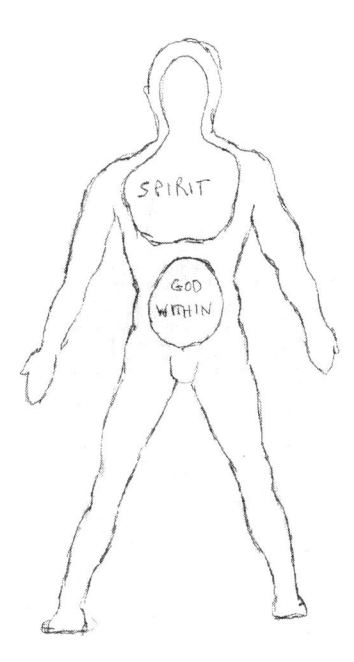

Fig. 2. The spirit in the heart and the god within

enthusiastic. (Ref.11b). It is omni-sexual but its activities and appetites are controlled by the spirit in the heart, (the essential self) and the head working in tandem, (which together make up the conscious mind) which is where the sexuality of the individual, hetero – homo – or bisexual is determined.

The GW can be roughly equated with the id in Freudian psychology but is much nicer and not so utterly selfish although its appetites are strong and very important.

Lyall Watson, in his book 'Beyond Supernature', tells of a brave anthropologist, visiting the Conibo people in the Peruvian Amazon, who was told that he would never understand them unless he drank a sacred drink made from 'ayahuasca', the 'soul vine'. This he did with the result that he had vivid and disturbing visions of crocodilian demons, soul boats and bird-headed people. He assumed that the imagery must be drawn from his own cultural background but was astonished to find that an old Amazonian shaman knew exactly what he had seen, and had had exactly the same experience as he had. This is easily explained by TMM. The personal unconscious of the anthropologist would be merged via BP with the local group unconscious of the Conibo so that when he drank the brew his unconscious mind would be flooded with material from the latter, which would then become manifest in his conscious mind.

In the same book Mr. Watson goes on to describe a case of multiple personality, Billy Milligan, who produced twenty-four characters one of which, Arthur, spoke with a British accent and demonstrated an extensive knowledge of physics, chemistry and medicine. He could also read and write fluent Arabic. Another gave his name as Ragen, spoke with a thick Slavic accent, expressed an interest in weapons and was skilled in karate. He proved that he could read, write and speak in flawless Serbo-Croat. All of which were well beyond the capability and opportunity of young Billy. This cannot be explained by unconscious telepathy but it can be explained as possession by an alien spirit or rather, spirits (Ref: 11a). The associated GW would remain unchanged but it is conceivable that there would be a collection of alien spirits, either in his heart or somewhere else in his body, or hovering around the subject awaiting their turn for occupancy. This observation is not actually part of TMM but it shows how a fundamental component of it, the acceptance that the spirit exists, can provide tools for the understanding and explanation of related matters. In one essay it clears up the mechanisms of Buddhist meditation, the vagueness of previous conceptions of the self, the 'hard problem' of consciousness, etc. as well as much, if not potentially all, of psi. It must be acknowledged, however, that I can't provide actual proof, as yet, in confirmation of many of my ideas, except that I do believe that the mere fact that one can mentally conceive perfect geometrical figures, especially in the context of the sticky psychoplasmatic background, itself constitutes a 'near proof'(see addenda 115) that the spirit exists, as does also the experiment conducted by Charles Tart (ref. 9a) and the events described by Dr. Kubler-Ross (ref. 13). It would be interesting to ask a person who had frequent OBEs to try to establish contact, during his next one, with a member of the AL, since he/she, if my conception of the AL is correct, would be effectively much closer to it when out of the body, although he would still be anchored in the body, by the GW which would still be rooted in his gut.

All I can do apart from this is paint, I think, a very convincing picture, and provide a strong, self-consistent argument, in support of my discoveries, although I have found material which amounts to a solid proof. (See addenda 115). Coming now to the end of this section I would like to make it clear that the god within is not an invention or idea of my own but a real, existing entity of whose presence I have gradually become aware. I understand, of course, that the idea of a god within must seem ridiculous and retro-grade to the average scientist and I, too, have had my doubts but a little thought on my quite astonishing experiences has always restored my belief, and it goes without saying that the god within is not - comical thought - to be worshipped, but simply accepted as the basic source of energy and feeling in the organism. It is the 'something inside that cannot be denied'

Finally, Is the GW omniscient and omnipotent ? We have seen that it transcends space and time and the laws of physics. Are its activities limited to and by the memories contained in the unconscious, that is, is its **knowledge** limited to the information stored in the PU? This question was provoked by the 'tribe' ghost experiment (Ref. 12) in which a group of experimenters sat round a table and tried to summon a spirit, in such a manner that its appearance could be attributed to the operations of the mind, rather than the paranormal. The answer is, NO! its knowledge is not limited to the memories in the unconscious. What is intuition? It is the power of the GW to create something brand new, something which did not exist until the GW performed its magic. As regards omnipotence, it is clear that its obligatory function of serving the spirit in the heart, imposes an immediate limitation on its powers in this direction, since it **must** obey the conscious mind. It is, indeed, the 'fire in the belly', **enthusiasm**, which fuels action.

❖ 2. The nonreplicable nature of psi experiments.
It is in the nature of the god within, in its status as the root and fulcrum of all living things, to mischievously deflect all attempts to analyse and classify its behaviour. It will not allow itself to be pinned down for dissection. Since it is the basic force behind all psi activity, this means that conclusive, predictable results of experimentation with psi will remain impossible. The god within will often fail to deliver the expected results in an experiment, and will play tricks on the experimenter when he is least expecting it. It is the agency that operates when, after a

conclusive series of psi experiments, it delivers exactly opposite, negative results when one attempts to demonstrate them to an audience. It is the 'Jokerman' in Bob Dylan's song, and there can be little doubt that the god within, in tandem with the spirit in the heart, as written, is the key to all psi phenomena. This proclivity of the mischievous GW to refuse to co-operate in the laboratory is the reason for the failure so far of the profession's efforts to present itself as a convincing and reliable scientific discipline. One could ask, 'what is the normal, day to day, function of the GW?' Or 'how does it help in the struggle for survival?' This is easily answered. Its primary function is to supply enthusiasm, love, lust, joy in activity or positive motivation to its possessor; that is - ***feeling***. This despite the fact that it can at testing times feel nervousness and fear.

All through the day the organism is busy looking after its own interests and the GW supplies the metaphysical energy for this, feelings of enjoyment, satisfaction, love, nervousness, whatever, and is a vital component (despite its location in the unconscious) in many conscious mental activities, such as judgement, belief, imagination, deciding and so on. The conscious mind could not exist without it.

❖ 3. The 'artlessness of youth'

When one is young one is joined on an unconscious level to lots of other people, young and old. This explains the fluid easiness, range and inventiveness of youthful expressiveness; one has access to many unconscious minds. As one grows older one 'individuates' and develops a personal philosophy which gradually isolates one from all but one's immediate and also aging social contacts, and as a result the pool of unconscious information that one has access to gradually diminishes. This manifests itself, moreover, as a reduction in the store of basic good feeling which animates the young, and which, in TMM is seen to be simply one's supply of psychoplasma. I can think of many examples of youthful expressiveness from my own youth but they all seem to be based around sex, which itself is significant since density of PP is related to sexuality, and when one is young one can, much of the time, think of little else.

❖ 4. The spirit in the heart (S) and the conscious mind.

The spirit in the heart, which can be equated with the ego in Freudian psychology, but which is a real, existing metaphysical entity whereas the ego is purely a way of thinking about self (Ref.10a), directs the conscious, volitional actions of the body and the activities of the conscious mind, as far as is permitted by the head.. It is the conscious, animating force of the organism and is the foundation of the conscious mind. The head can be partially equated with the super-ego and the familiar balance between head and heart, but is also responsible, of course, for intellectual and sensory activities, and regulation of some bodily functions. The body derives its physical energy from chemicals obtained by digestion of food in the stomach and the gut. Its metaphysical energy, strength, vitality (origin of 'gut reaction') and enthusiasm, its status as living matter, is derived from the presence in the gut of the god within working together with the spirit in the heart. The spirit directs energy and vitality from the god within into the appropriate parts of the body, muscles and organs in general. (Precisely how this happens I am not quite clear as yet). The spirit, which, though based in the heart, is present throughout the upper body including the head, must send an electro-chemical signal to a given area in the brain, or manipulate PP already existing there by virtue of the fact that BP is generated in all of the body's cells, and the brain then sends an electro-chemical message through a nerve to actuate the relevant muscle. There would also be a flow of information towards the spirit which would be registered as perceptions

The human spirit and that of the higher animals originates, as stated, in the heart, that is, the chest cavity, and extends throughout the chest, neck and head, and, to a lesser extent, all of the body. It is, I believe, simply a presence, an awareness, which displays intentionality or purpose, is possibly immortal, and leaves the body at death to join other departed souls in the afterlife, which has, without a shadow of doubt, been found, by me, to exist. I shall describe it later as far as is possible. I don't know precisely how the spirit is connected with the physiology of the body but it is clear that one's centre of awareness in any action is usually located in the heart or thorax, which is the locus of the essential self (or spirit), the basis of the conscious mind and the initiator of action. The spirit is the origin of consciousness, courage, motivation, initiative, conscious, directed, voluntary thought and willpower and is the seat of perception. (ref. 2).

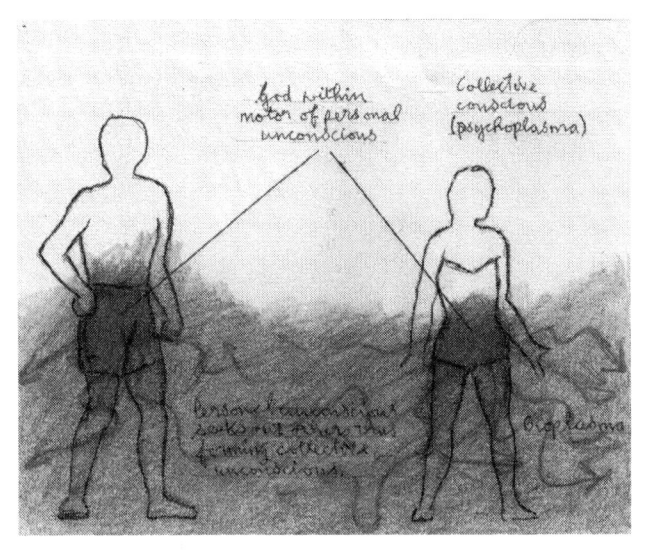

Fig. 3 The gods within reach out to join with other
gods in other people and higher animals.
(Arrows are diagrammatic and have no real existence)

The senses respond to the stimulae from the environment, electro-chemical messages are sent to the brain, where they are processed, but the actual perceptions, the colour, sound, smell, touch and taste, the qualia, are metaphysical things and they are realised by the spirit, at the interface of body and spirit, which is present in the head and upper body. That is, there can be no true perception there until the electro-chemical information makes contact with the spirit. I base my contention that the spirit is the origin of consciousness, mainly on the fact that in out of the body experiences (OBEs), the disembodied spirit stands near or floats above the body and sees it as it is - unconscious. But the consciousness is still there in the spirit, which is wide awake and observing. The OBE is a fairly common occurrence and has been well-documented to show that it is a genuine case of the spirit, or some part of the individual which leaves the body (ref. 9a). There are many more well documented cases but one will suffice for my purpose. There is no doubt in my mind that the spirit does exist and is not a poetic or literary fiction, and the same goes for the afterlife which is the destination of departed souls or spirits. My certainty about the existence of these things is based not only on reliable evidence reported from experiment but also the result of experiences I had which brought me very close to dying. On both occasions I could feel my spirit fading out of my chest and heart and knew that if I didn't make the effort I would pass away. Luckily I managed to hang on.

I used to believe that the spirit was an essence but now I feel it may just be an invisible but living and animating presence or awareness. Descartes thought it was a nebulous substance. It seems, to me at any rate, in view of my findings, so OBVIOUS that a metaphysical explanation, as evidenced by the OBE, must be allowed, that the idea of looking for a 'consciousness gene' or subatomic 'tubule' as the origin of consciousness seems, if not comical, dubious at least, and is clearly evidence of original thought being strangled by the dogma of reductive materialism.

It is clear that the entity that rises out of the body during the OBE is capable of seeing without eyes (ref. 13) and I'm pretty sure that it can hear too and maybe sense touch, taste and smell. I am sure that all perception originates in the spirit (the spirit is the person, the bedrock of perception and therefore, with the GW, ultimate reality) and in view of this it is conceivable that the afterlife is just as sensually satisfying a place as the present one, since it may be that its occupants can generate desired perceptions at will. More realistically, however, it would seem more likely that they engage in more purely intellectual pursuits. Who knows? An interesting experiment would be to take two people, A and B, who have frequent OBEs and to ask them to try to find and communicate with each other while out of the body, and also to try going into the past or the future, a possibility which has been confirmed for me by accounts in 'The Holographic Universe' (Ref.17) although the future excursion was limited to a few days only. It would also be interesting to try, while in an OBE, making contact with a member of the afterlife.

❖ 5. The dual nature of reality.

The TMM is dualistic as to its essential philosophy; that is, there is both mind and matter. However it differs from other philosophies insofar as it consists of one physical and one metaphysical component which are joined, weakly or strongly, depending on circumstances, by material bioplasma (BP) which is secreted, I think, in the cells of the living organism. It is the physical quality of BP that enables PK, and its spiritual or metaphysical component enables telepathy and much if not all other psi.

This essentially dualistic reality has three aspects:

(a) The reality of matter
(b) The reality of perception.
(c) The reality of the divine..

A description of these aspects follows:

(a) The reality of matter

The 'reality of matter' implies that the body and its environment are real. This agrees with Kant's concept of 'noumena' – things which exist in themselves. It also agrees with Russell's vision that a table as seen by Tom, Dick and Harry is described by each of them in much the same way, which tends to verify its reality. That is, there *is* something 'out there', which is the cause of the sense impressions described by the philosopher David Hume. This could also apply to bioplasma which, I believe, is created in the cells of the body, but which, so far as I know, has been explained by no-one but myself, although its existence must be known of by many, Kate Bush included. I mention this because of her song in which she mentions 'sticky stuff" (psychoplasma). (see ref.7)

The reality of Bioplasma is fundamental to the TMM so it is safe to claim that reality is, partly at least, material rather than ideal, and indeed the 'mind over matter' aspect of TMM works precisely because bioplasma animated by non-material spirit and the god within, is a substance. There is a religious philosophy called 'foundationalism' which seeks to assert that all things revert or relate to an uncaused cause which is felt or self-evident and is not open to logical exegesis. It occurred to me that the GW satisfies this requirement but on closer analysis it became apparent that the TMM requires, of necessity, the additional existence of the spirit, bioplasma and the afterlife. These things are not parts of a constructed philosophy but actually exist.

(b) The reality of perception

My current conception of perceptual reality stemmed originally from the idea that the most real existent thing of which the average person can be aware, is extreme, excruciating pain which is identified as, above all, a feeling. This lead, in part, to the formulation of the concept, 'the Primacy of Feeling' (see chap.2 item 10). If someone is suffering agonizing toothache there is no doubt to him that the pain exists. It is real, not imaginary or illusory. It is a feeling and one can conclude therefore, that feeling - all feeling oddly enough (see to feel is to know chap.3 items 6 and 7) is real; it is perceptual and percepts are real; but the spirit, which does the perceiving, represents ultimate reality than which nothing is more real. (It is important to note here that percepts are registered by the spirit, and that all and any incoming information reaching the spirit, the basis of personality, is registered as a percept). The taste of honey, warmth of the sun, the blue of the cornflower and aroma of lavender on a warm, Spring day; all these are real. (qualia in the words of the neuroscientist). When I touch the cool, hard table it is not the table that is real but the sensation, the percept, or, to be more exact, the percept is stronger and more certainly existent than the object. By the same reasoning I can conclude that the body is less certainly real than one's perception of it.

The object then, let's say a table, may be real but the reality of the percept is stronger and more certain and percepts can exist independently of the object, (hallucinations) which may not exist at all, although these latter, hallucinations, would in all cases, I think, be caused by erroneous or spontaneous electro-chemical messages sent to the brain, which includes drug-induced hallucinations, and means that there *is* a certain element of reality associated with them.

It is clear that the TMM embodies a triplicity, body, spirit and the god within, since it is a fact that both the spirit and the god within are real, existent things, though the body, also real, is less permanent and enduring than the other two. It is possible, if one accepts the buddhist belief, that the spirit is finally absorbed into 'the great spirit' thereby extinguishing identity and individuality, but one has to bear in mind here, that my experiences suggest that a spirit can endure for millennia, (Heraclitus, Plato and Pythagoras are still spiritually extant as shown by my contact with them (Chap.8 item 6 no.58. 59.60.) and even millions of years; this is shown by my contact with the hominid which died seven million years ago but is still telepathically accessible, and therefore in possession of its identity and faculties.(chap.8 item 6 no. 26) This, it seems to me, puts in question the buddhist conception of 'the Great

Spirit' and also raises the question of 'if the spirit doesn't join the Great Spirit what does happen to it?' Perhaps it *is* immortal. This is a very difficult thing to imagine. The conscious mind necessarily looks for an end to things in time or space, unless by accepting that there are no ends, it somehow grows to encompass these ideas. Elswhere in the essay I claim that there's no such thing as time, especially in the afterlife, in which I was in agreement with Einstein and Kurt Gödel. This requires more thought and must be dealt with in my next book.

At this point it would seem advisable to remember the fact that the unconscious mind, motivated by the GW, (as opposed to the conscious mind), and in communication with the other six billion unconscious minds of the peoples of the earth, thus forming the collective unconscious, is effectively infinite and would have no difficulty in accommodating the concept of the immortality of the soul, or the intellectual impenetrability of quantum mechanics, (which is, anyway, despite the fact that nobody really understands it, already accepted by physicists as orthodox science, especially since it has generated such a powerful technology). Therefore, since infinity of space and time is easily accepted by the infinite CU, it must be true and we must be content with that assumption.

Since first becoming interested in defining reality I have modified my picture slightly; i.e. the spirit, in tandem with the head (which together form the conscious mind), is ultimately a stronger thing than sensation alone. Logically it must be since it is the spirit that does the perceiving, that is, sensation cannot exist independently of the spirit, but the spirit can exist independently of sensation in the body. If proof of this superior power of mind is required one has only to remember the TV pictures of Buddhist monks in the nineteen sixties, demonstrating against the Vietnam war, seated in the lotus position, being drenched with petrol, setting themselves on fire and, without moving a muscle or uttering a sound, calmly burning to death. Such dedicated buddhists train themselves for years with the aid of a 'mantra', to such a degree that finally they can turn off unwanted bodily sensations and associated percepts. There is at least one mental illness in which the mind blocks bodily sensations, and I myself can testify to this, having once met a lady who habitually stubbed out her cigarette on the back of her hand.

She assured me that she felt nothing. Therefore, modifying an earlier statement, and in view of this and the self-immolating buddhist monks, we are forced to the conclusion that the most real existent thing is the spirit and its attendant GW.

(c) The reality of the divine

Both the conscious and the unconscious minds have at their centre a metaphysical agency, (1) the spirit and (2) the god within, and both of these agencies have access to memories located in differing parts of the body. These two entities can be diametrically opposed in nature; the spirit and the conscious mind embody the character or personality of the subject which may be warm, loving, lusty, free and easy, or cold, prim and proper, even prudish or fuddy-duddy, whereas the god within is always the same, it doesn't change its basic nature which can be thought of as free and easy, jocular, warm, gregarious, sensual, enthusiastic, displaying an amoral love of people, things, activities, whether the subject be man or woman. It can be seen therefore, that reality is ultimately metaphysical, that is, the strongest and most enduring aspects of reality are the god within and the spirit. Even if we go back to Tom, Dick, Harry and Bertrand Russell agreeing about the reality of a table, they were all employing the metaphysical agencies of the god within and the spirit, when making their judgements. This, however, does not invalidate the fact that the table, or something related to their sense impressions, must have been there. This makes it possible to claim that the human being is basically tripartite, that is, material, perceptual and divine, and ultimately the perceptual and divine, both metaphysical, are the strongest and most enduring ones. Thus we have an effectively dualistic reality; basically a platonic world-view.

In chapter 5 item 6 (the duality of imagination) I describe experiences which introduced me to the ideal forms described by Plato. Although it is not necessary, it might be helpful to the reader to skip forward to read them since they relate to the reality of the spirit, and can almost be considered to constitute a proof of its existence.

Graph 2 represents the density of PP as it would be for five adjacent subjects. The density would be '1' in the heart of each subject tapering off to lower levels with increase of distance from the subject. Maximum density '1' in the heart of each subject would need to stay at this level, or at least well above the basal density, to maintain life and this would be done through the cooperation of the GW and the spirit working together in the body.

❖ 6. The origin and nature of thought
This heading can be divided into two subsections.

(a) Voluntary thought.
(b) Involuntary thought.

In this context 'thought' signifies any mental activity i.e. imagination, understanding, calculation, remembering, analyzing, classifying, attention-directing, thinking in words, cognition, etc. The first (a) of these kinds of thought originates mainly in the heart which is the origin of conscious, directed willpower, courage, initiative, the ability to manipulate PP. Voluntary thought operates within the mass of PP in the chest cavity and in the head where PP is more concentrated and therefore more susceptible to manipulation than in the heart itself. Much if not most voluntary mental activity, takes place in the head and continues outside the head in the surrounding space to a distance of one or more metres.

The second (b) of these kinds of thought, can occur anywhere in the body and is, I think, sometimes the result of an accidental unconscious connection with a neighbouring unconscious mind, for instance if David is waiting in a room with Bill they will be automatically sharing BP since their unconscious minds are material and to a large extent share the same space. It would seem likely that they would also be strongly connected on a consciouslevel, since their conscious minds, their PP, would appear to be automatically mixing. This though, I think would depend on whether david and Bill were acquainted or not. If they were not and were not engaged in conversation, the PP, instead of mixing would draw back to rise in what I can only describe as a cool, self-contained tower, thus minimizing contact with each other's PP, and thereby retaining its 'mental privacy'. If they were acquainted they would probably chat about this or that and there would be a certain flow of PP back and forth between them. I should report here that the image of a silent tower of

PP formed quite naturally in my mind as did the 'funnel' to the ISS described later. It was not a result of mental effort and felt as though the unconscious was revealing the truth. It *felt* right, and has also cleared up a lingering problem, i.e. ever since I started this essay, working alone without consultation or discussion, I have been at risk of having my ideas stolen by accidental telepathic connections, and this discovery has informed me that the body plus its PP, automatically puts up defences against telepathic theft. I have in my possession, an envelope full of rival psi theories none of which I can read since the moment I attempt to, I am in TP contact with their authors and feel my ideas being stolen. It could be suggested that I could *share* my ideas with the 'telepathic eavesdroppers' but I don't need to. I know my theory is correct and I would be giving vastly more than I got.

Involuntary thought arises in the flesh of the body, the cells presumably, and probably all of the body's constituents, bone, liver, connective tissue, hair, and tooth enamel and is totally undirected. It consists of feelings of mass, body consciousness, substance, images, words. My observation of involuntary verbal thought has occurred on at least two occasions, (a) thirty years ago during my lonely period. I was sitting on my bed grasping a roll of fat on my slightly overweight abdomen, and suddenly the words, "this stuff thinks!" came into my mind.. I took this as the exact truth – flesh thinks – which has since been confirmed - an addition to the other messages I've received from what I believed, correctly I think, to be the the collective unconscious. My second observation came a couple of days ago. I had gone upstairs to my flat and, entering the kitchen was aware of my head and the space and light in the room. Five or six seconds later I became aware of a mass of small, quiet words originating, rising, within my left thigh, abdomen and gluteus maximus. The first thing that the little voice said was, "I'm in the kitchen" and I thought at first that it was the GW doing the thinking, but now, reading it over again, I can see that it was simply myself, in my conscious mind, making an observation about my body's topographical location . As can be imagined - I would hope - by my reader, I have had some extremely confusing moments on my long and difficult journey back to reality and ultimate knowledge (which now I am sure I possess), about the nature and mechanisms of psi.

While all this was happening 'I', in my conscious mind, simply observed the arrival of these involuntary thoughts without interfering. This kind of thought must be happening all the time as a background to the individual's mental activity. There were, however, in my case, the recurring thoughts such as 'all gods are good but some serve evil ends', 'the spirit informs the flesh' and 'the inner truth is greater' which came from I know not where but whose truth I accepted.

❖ 7. Hume, metaphysics and the self.

David Hume, the philosopher, when he was 'looking inside himself', and failing, to find something which he could legitimately call the self, was actually using this entity, the 'self' or spirit, in its developed form as the conscious mind, to look for itself like a dog chasing its tail. It is interesting to observe that despite his rejection of all things metaphysical, and his position as a leading intellect in the Enlightenment, he did not believe that human activity was based on or driven by reason, but by 'the passions' or feeling. These things, passion and feeling, are essentially metaphysical in nature and originate with the god within and the spirit in the heart. The self, I believe, can be thought of as having three components:

1. The spirit or 'essential self' which animates the conscious mind and, via the brain, the body.
2. The 'acquired self', the person with experiences, opinions, likes and dislikes, feelings, ideas, ambitions, skills, hopes and fears, 'the collection of habits that sits down to breakfast' to quote Yeats, (accurately I hope!).
3. The somatic self, the body.

I think the spirit must grow in step with the somatic self and the acquired self to incorporate things learned; the spirit of a baby must change with advancing years as it becomes that of an old man, although it retains its identity. (There can be only one body per spirit; I call this fact 'the individuality of the spirit'[IOTS]). I am quite certain that the spirit goes to the afterlife at death and I am pretty sure that the whole self or mind does so, because in my contacts with individuals there, I always have an unmistakable feeling of their identity, and I sometimes get verbal messages. Hume's rejection of all things mystical or metaphysical was a characteristic of the times. He lived during the Age of Reason (The Enlightenment) which preceded the Romantics and tried to reduce nature to mechanisms described by reason and mathematics including belief in God, (see deism chapter 8) and substituting instead, the 'clockwork universe' in which men were enabled to manipulate matter and create such things as clocks and watches, steam engines and railway trains. I should at this point assure the reader that I myself do not believe in a creator God or supreme being, although I am quite certain that nature and all living things, are populated and joined by omnipresent divinities and intelligences, greater or smaller.

❖ 8. The collective unconscious (CU)

The BPE connects all life on Earth and is, in addition to other things, the material embodiment of Jung's Collective Unconscious and its active, motivating components are the 6 billion human gods within and those of the higher animals. Ultimately every speck of life plays some part in the CU, but micro-organisms would operate mainly in and around other, larger organisms.

Large masses such as plankton, mosses covering a hillside or large collections of micro-organisms and bacteria would be joined by a thin, tenuous BP. If one goes back in imagination to the dawn of life on Earth when these were the only life-forms present, one can imagine a general mantle of tenuous 'godliness' which covered and permeated them, for the gods must always be present in some form I think. And before there was any life here, only an incandescent, cooling planet? Where were the gods? Who knows? Perhaps on other planets in other star-systems, for gods I think, don't die. Perhaps space itself is inhabited by gods not actually connected with any material body, but waiting for a suitable situation or opportunity to arrive. Space as described by the physicists is a subtle foam in which subatomic particles come into and out of existence. This is interesting because the gods must come into contact with these phenomena if they really do exist in free space outside the body, and may even share a causal relationship with them.

❖ 9. The collective conscious (CC).

There exists, complementary to Jung's Collective Unconscious, the 'Collective Conscious", which operates, among other things, in conscious telepathy and other paranormal phenomena between humans, animals, insect and other life, and vegetation. Qualifying this I see now that it must unite all living matter on the planet as does also the BPE (the CU) on an unconscious level, but which is much denser than the CC.

The CC is composed of psychoplasma which is the material of the conscious mind, and it covers the entire surface of the planet including the deserts and poles where it is less plentiful and probably drops to basal density, due to absence of life at these places. It is probably fairly dense in the seas and oceans. Its density varies from very thick in crowded pubs, suburban homes and concert halls, shoals of fish and flocks of birds, to thin and tenuous in less populous regions as explained a few lines ago. PP is a substance which is made of BP and spirit (S) and I think it's probable that these mix in proportions which will vary according to the locality and circumstances giving, ultimately, a varying density and quality of the substance, though there would be a basal density below which PP would not fall. The CC is essentially the sum total

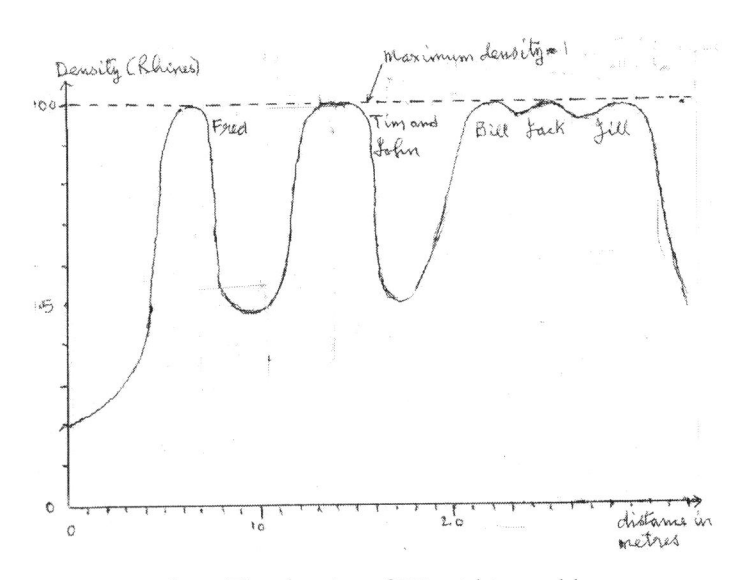

Graph 2: The density of PP within and between
groups of people versus distance.
(basal density not shown)

of all existing conscious minds plus a 'no man's land' of PP which is free and accessible to any conscious mind. At times the CC will be calm and at such times it will hold the subject's mind in a calm, light, limpid volume of PP. At other times and places it may trap the subject in a blob of thick PP hampering his movements which may become sluggish and lethargic and slow his mental capacities. (See Chap.7 items 8and 9). On occasions where the subject is 'on the crest of a wave', this state would manifest itself as a feeling of joyful lightness of heart and spirit.

PP is present in large densities in and around the head, neck, shoulders, arms and to a lesser extent the lower body and extends into the surrounding space and the CC. It is the connecting, rarefied fluid medium posited by Arthur Koestler, by which conscious telepathy operates. (There is also unconscious telepathy in which the proportion of BP to S is higher, I think, than in the conscious mind's PP). This variation of the ratio of BP to spirit with altitude is included in fig.4

❖ 10. The primacy of feeling (POF).

The primacy of feeling is a universal principle and it operates in all living systems. Feeling is real and its dictionary definition is 'capacity to experience the sense of touch, a belief or opinion'. Logic is a tool. Feeling affects every one in the same way. One can locate the truth (about some matter or person) via feeling (to feel is to know), e.g. one's next door neighbour is *felt* to be a friend, or by logic and the machinery of reason, e.g. the existence of Neptune was *predicted* by the use of mathematics. (See chap. 3. item 6. and addenda 107)

In POF the operative mechanism involves feeling, and always involves maintaining a balance between feeling and logic. Feelings, whether physical or mental, are essentially vital and life-centred, and their possessor may suffer or rejoice at the hands of cold, unfeeling logic, which is a tool. Feeling can be classified under three headings:

(a) purely physical: Pain, aches, tiredness, muscular strength or weakness, well-fed, hungry, thirsty, energetic, loose limbed and agile, athletic.
(b) purely mental: Consciousness, despair, hope, alertness, clarity, vagueness, sharpness, the feeling of knowing, doubt, uncertainty, certainty, belief.

(c) Emotional: Apathy, enthusiasm, joy, anger, pride, impatience, greed, triumph, joie-de-vivre, lightheartedness, assurance, confidence, sadness. These feelings are accessible to all conscious minds, and it is certain that a minimal degree of consciousness operates even in vegetal life as implied a few lines ago (ref. 6).

It is a fact of course that things are not quite so simple e.g. the differently classified feelings can interact so that feeling 'x' affects or causes feeling 'y', e.g. I stub my toe on the projecting table leg, which causes physical feelings of pain which in turn cause mental or emotional feelings of anger or annoyance.

The common characteristic of these classes is, of course, their status and description as *feelings*, and the principle of POF does not distinguish between types of feeling, it simply recognises them as being in opposition to or in support of whatever logical procedure is in process. This fact should be remembered while reading the examples which follow a little later. The human spirit and that of the higher animals originates, as stated, in the heart, that is, the chest cavity, and extends throughout the chest, neck and head, and, to a lesser extent, all of the body. It is, I believe, simply a presence or a region of awareness or consciousness. Descartes thought it was a 'nebulous substance'.

If the POF is not observed in a given activity or circumstance e.g. local or national government, the result will tend towards over-use of the intellectual or organising side of government, with consequent centrification, runaway bureaucracy and harm to the forgotten electorate, whose feelings will be trampled on. Such a situation could be illustrated by a hypothetical office manager who, swayed by a love of order and logic, and with the best of intentions, introduces more and more rules and logical procedures into the office, all designed to speed up and simplify its operations but resulting in bad feeling and dissatisfaction in the employees who rebel against all the new paperwork they have to complete and the unnatural, intellectually conceived procedures they must observe, instead of being allowed to follow routines which had grown naturally. This state of affairs in an individual person would produce someone who was no longer 'in touch with his feelings' i.e. clumsy, insensitive and socially inept or tactless. In this example POF was violated.

- Why is POF important? Because it is absolutely fundamental in any search for truth, since it is also the basis of judgement and belief. Feeling will usually reveal the truth of a situation where logic may be deceptive or ambiguous. The opposite, too, applies of course. Logic may sort out the situation where the feeling is weak or mixed with other, conflicting ones.

General Facts:

1. The Primacy of Feeling is based on a general principle; i.e. if feeling and logic conflict logic must give way unless the feeling can be adjusted to accommodate the logic, and achieve a state of balance. If not feeling must prevail and ignoring it will usually have regrettable consequences. This applies on a personal and a universal scale, that is, stated in demotic: "whoever you are, wherever you are, whatever you are doing, it's got to *feel* right. If it doesn't something is wrong." One does, of course, sometimes have to do things which ***don't*** feel very nice e.g. clearing up one's ailing child's vomit, but the fact that one *knows* that it has to be done legitimizes or cancels out the bad feeling. The conflict in POF is between feeling and logic, not feeling and knowledge. Knowledge, if true, is unassailable but logic is a means of acquiring knowledge and its conclusions depend on its premises, assumptions and a sharp eye for mistakes and wanderigs from the point.

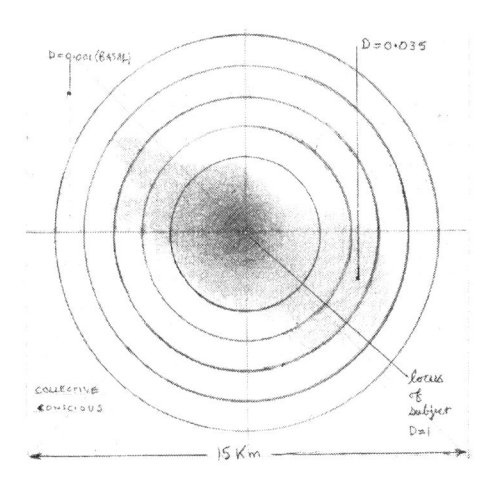

Diag. 2. Plan view of a large material conscious mind (shown unconnected with other minds). Personal Unconscious – not shown – would lie behind the conscious one

2. Thoughts are tiny feelings. That is, a thought is a discrete larger or smaller quantity of PP which, in some form contains feeling(s) and, possibly though not necessarily, images, ideas and words, or any one or any combination of them. This agrees with conclusions written elsewhere, to the effect that I believe that PP contains information. It also agrees with ideas expressed elsewhere in the essay i.e. in any telepathic communication a discrete, larger or smaller quantity of PP flows between the participants. In this flow, the discrete volume of PP may not have to reach its objective (the mind of the receiver – which may be thousands of miles away) in order to be registered, but may simply move over a fraction of the distance separating it from the sender so that its effect

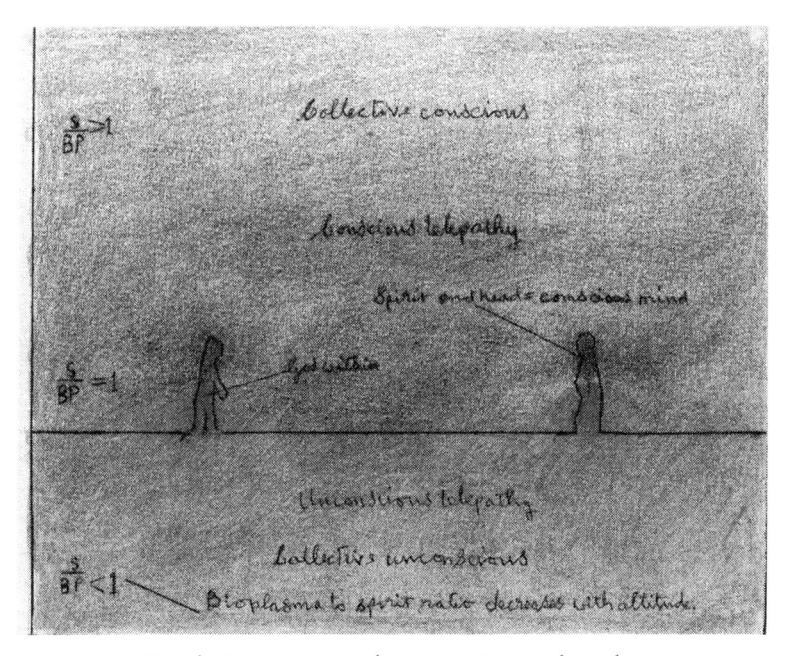

Fig. 4 Conscious and unconscious telepathy.

is registered at the receiver. This can be illustrated in the following way; a three-foot long tube with a 'bore' of ½ inch diameter is filled with plastic balls of very slightly less than ½ inch diameter, so that they can move. If one now pokes one's little finger into one end, two or three balls will move at the other end so that an effect will be transmitted but the actual balls near the sender will move only a short distance.

3. All knowledge comes from feeling (Leonardo) including knowledge of oneself and one's existence, therefore D'escartes was really saying 'I *feel* therefore I am. (Ref. 5)
4. Everything that lives is holy. (William Blake)
5. The organ of knowing or cognition, and reason and logic, in tandem with the heart, is the head.
6. The organ of perception is the spirit, a metaphysical entity.
7. The sense of self and the personality are based in the head and heart.
8. Feeling and instinct are based in the gut.
9. Intuition operates by means of head, heart and gut working together.

❖ 11. Illustrations and assertions supporting the concept of POF.
- The following example actually happened as described:

I was checking my bank statement to see whether a certain cheque to 'Argos' had been cashed. It was payment for a TV, not yet delivered, and I wanted to cancel the order. Sure enough, there it was on the statement, so I concluded, of course, that it had been cashed, and should be refunded to me. Then I got a silly desire to check in my chequebook, (evidence of a certain habitual pedanticism in the execution of my affairs, and a regrettable waste of time I thought), to see the actual cheque stub. So then, calling myself all sorts of idiot, I slowly and indecisively opened the book, thumbed back to the cheque date and, to my surprise, discovered a stub to the same firm for the same amount but for a different object – an electronic dictionary – so the encashed status of the cheque was perfectly in order, and no refund was due. My unease was well-founded! Something, (the god within at the seat of feeling), knew – or felt - that I was making a mistake and would not let me relax until I'd checked! (POF observed).

- If a person is presented with a theory based simply on logic (mathematics is basically logic as shown by Whitehead and Russell), he will accept it as long as it doesn't stretch the god within too far as it adjusts to accommodate the changing states in the conscious mind, (of which it is aware via the BP component of PP) which executes the logical manoevres of assimilating the theory.

For example; if a mathematician were to advance a conclusive proof that the creationist theory of the origin of species was true and Darwin's Theory of Evolution by Natural Selection was wrong, one would, sensibly, disbelieve (although some diehard types might claim that maths is infallible, and would therefore accept the proof. [see addendum 112])

All this represents a very interesting hypothesis as to the mechanism of belief and judgement, in which the GW plays a crucial role. I have stated elsewhere in the essay that the GW never tells a lie and never makes a mistake. It is always right in its judgements which are delivered either from communications with neighbouring intelligences, (gods within), or from a mysterious region beyond space and time which is inaccessible to science or any other human institution. This hypothesis is described more clearly in the following paragraph.

❖ 12. The mechanism of judgement and belief.
We have established that the god within has access to the memories stored in the unconscious of which it is the motivating force, but also has access through unconscious telepathy to those of its neighbouring PUs. These memory traces refer to things in the past but have existence in the present. The GW, however, also operates beyond space and time. It has access to things which used to exist but no longer do, and presumably things which will exist. Let's think now of what happens when the subject 'believes' something. We established in the preceding paragraph and elsewhere in the essay, that the GW has access to 'anywhere and anywhen' and is therefore omniscient, and always right in its atitude to the world. It *feels*, hence 'the primacy of feeling', and adjusts to whatever message it is getting, via the BP component in PP, from the conscious mind, which is where reasoning and logic take place. The god within never makes a mistake. It just feels. First the conscious mind apprehends the entity which is to be judged and holds it within the field of the GW. Then the conscious mind begins a dance with the god within and, holding the entity before it, the two merge and become one, at which point the judgment is made or the belief established.

It is interesting to consider the existence and role of intuition. Intuition, though not infallible as a way of acquiring knowledge, is an essential ingredient in matters of creation, whether creating new poems, paintings, inventions or scientific theories. Intuition presents us with a brand new thing – not just recombinations of already existing things – which would mean there is a limit to creation – where the ***last new thing possible*** could be pulled out of the bag, leaving the bag empty, creativity at an end and logic 'king of the hill'. Intuition is an elusive, mysterious and unpredictable thing and shares these qualities with the god within, wherein it originates. On these grounds, therefore, we must accept that the GW is infinite, unpredictable, omniscient though not omnipotent, and has access to anywhere in the cosmos. It will, I think, in league with others of its number, prove eventually, to be our route to interstellar and even intergalactic space travel. (see chap.5 item 13).

As to the kinds of feelings to which the GW is susceptible, let's list a few: First of all, enthusiasm, love, appetites, friendliness, mirth, fear, nervousness, (butterflies in the stomach) warm feelings, lust, etc., and the heart, pride, anger, love, equanimity, patience, sympathy, joy, depression, elation, and similar feelings related to willpower, idealism, the pursuit of goals, fixity of purpose and intentionality.

Another reason why feeling is prime is because among the more advanced animals from mice upwards, and also I believe, smaller life forms, down even to bacteria and other micro-organisms, feeling is basic to being alive. When one displays an advanced computer or artificial intelligence system as having a comparable computing capacity to a human, the first question asked is, 'can it feel?' And of course it can't. It is inanimate matter.

There was a program on television a few months ago about infinities and similar large numbers. Unfortunately I failed to make a note of the date and channel. It dealt with such concepts as the bevy of monkeys at their typewriters who, given enough time, according to the laws of chance would produce the sonnets of Shakespeare. The program claimed that a mathematician had calculated that to produce one complete line of a single sonnet would take the entire span of history since the occurrence of the 'big bang'. The big bang itself, according to the physicist Roger Penrose, had happened although the chances of it's doing so were only one in $(10^{10})^{123}$. This fact, if it is one, does not surprise me in the least. I have deep-seated reservations about the big bang's veracity and in fact it represents, for me, a perfect example of the law of the Primacy of Feeling being violated. I'm afraid that I find, along with others, that I simply cannot accept that something as vast as this universe of ours could have started as a dimensionless point of infinite mass expanding, along with the expanding dimensions of spacetime, to its present, still increasing dimensions. The expanding dimension of time too, I find unconvincing especially since I, again in company with others, including Kurt Gödel and Albert Einstein, don't believe that there is any objective something we can call 'time', but only change. Add to all this the presence of the multitude of gods, intelligences and spirits that inform the TMM – whose presence, admitted, is still to be proven, and the probability of their being eternal and the Big Bang could be seen to offend commonsense. It is interesting, in passing, to observe that unlike Relativity and Quantum Mechanics, the Big Bang has produced no useful technology as yet, which adds fuel to the arguments against it.

A parapsychological technology is already in evidence, in the form of numerous water, coal and oil diviners, and the much maligned, though very often fraudulent, mediums who contact the afterlife. My own contribution to this technology, an invention, which has grown from my theory, if it works, should make it possible to amplify the residual, low-level paranormal powers which every one of us possesses, and make it possible to easily locate lost objects or missing persons and even loosely predict the future.

The potential technological advance which this will represent, if it works, can be seen as of the same order as that between renaissance man using a telescope and 21st century man using his PC. Here, humankind hardware-wise, discounting the divining wand and pendulum etc., will step over the gap from physical science and into metaphysical science.

I now want to bring forward The Big Bang as an example of feeling being offended by deceiving logic. The big bang presumably, relies very largely on mathematics plus imagination for its development and explanation. Mathematics, as we have seen, is essentially logic. We have here then a perfect case of a reality created by a mainly logical argument, if one discounts Hubble's red shift and the discovery by Penzias and Wilson in 1965 of the background microwave radiation, which tend to support the idea of the big bang. This reality though, speaking for myself at least, and numerous dissenting physicists too, as it happens, feeling and intuition simply cannot accommodate. Something balks. It doesn't seem possible, whereas Einstein's concept of 'time dilation', though weird, was easily assimilated when clearly explained, and can be shown to apply by any sixth-former with the aid of pencil and paper.

The law of the Primacy of Feeling is fundamental to all science and indeed to all human life. It has to be since it is based in the fact of our being human and primarily and inescapably living by and through feeling(s) rather than logic; computers work by logic but they don't feel, and they can't create. That requires intuition. .(See addenda. 107 and 108).

Einstein, the profoundest thinker of twentieth century physics, had a deep mistrust of mathematics and was wary of being mathematically led by the nose into indefensible and unjustifiable areas; he was presumably deterred from such actions by feelings of uncertainty, dissatisfaction and disbelief (POF).

Whatever the truth vis-à-vis the big bang, although I cannot prove it as yet, there is no doubt that the gods, were, are and always will be, ultimate forces underlying reality. (see EPR paradox chap. 5 item 13 and ref.16).

Another example of conflict between feeling and logic and therefore an example of the truth of the POF is thesupposed fact that it is possible that there exists, at some enormously great distance from the here and now, another version of my own self, a duplicate of the constellation of atoms and molecules that go to make up Philip Hodgetts, President Obama or the reader him/herself. This is clearly, as was explained in a TV program of which,unfortunately, I did not note the details, another more extreme case of the typing monkeys effect. This was presented as fact on the program if my memory serves me well, and I was reading the same ideas, again presented as fact in a well-known scientific periodical, just a few days ago. I do, of course, understand the proposition perfectly but something deep inside strongly objects; there is only one me and that is the one sitting here who ***feels*** and the POF clearly indicates the absence of the other me which is a mathematical/logical construct. I ***know*** there is not another me because if there were I would feel it – and it would feel extremely strange. It also violates my principle of the individuality of the spirit, and such an argument must be dismissed as nonsense.

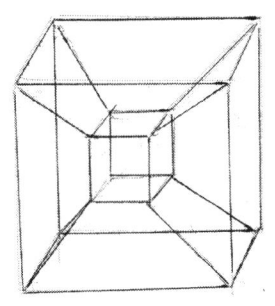

Fig. 5. The tesseract or hypercube

Physics contains several other concepts which, to my mind, POF has helped to cast doubt on, e.g. so-called 'additional spatial dimensions'. For instance the 4-dimensional cube the 'hypercube' or 'tesseract' (fig.5). This can be drawn and symbolically represented on paper in an easily comprehended diagram, but there is no way that it can be imagined or 'lived-in' in 4-dimensional space, and the multiple 'tiny, curled-up' dimensions that some physicists propose as true, are, I suspect, of the same nature. I believe also that 'parallel universes' is a fallacious concept, since it is, in essence, a variation of the quantum 'many worlds theory' which I know to be untrue since it, too, violates my principle of the Individuality of the Spirit (IOTS). For example; there is only one Jesus (now in the afterlife) and no re-incarnations of him, despite the claims of numerous deluded individuals, and the same goes for Beethoven and Julius Caesar. As to the many miracles supposedly performed by Jesus, I have little doubt that they are all quite true. (see addenda 101) i.e. walking on the water, changing water to wine, feeding of 5,000, the raising of Lazarus and the casting out of demons (curing mental illness), and even controlling the weather.

The possibility of the existence of higher dimensions is sometimes illustrated by referring to the inhabitant of 'Flatland', a place invented in a social satire by Edwin. A. Abbot in 1888. It is claimed that such an inhabitant would have no conception of a three dimensional world because his/her experience would make it impossible to imagineextra dimensions. This is clearly nonsense because such a hypothetical being could not possibly exist without the third dimension to give it volume, except in the sense of Plato's metaphysical forms which have existence only in the mind, and using an impossible concept as the basic premise in a logical argument is obviously wrong. One could, of course, point to a drawing of the 'flat being' and ask, "what is that?", and one would answer, "a drawing in three dimensions – length, breadth, and the thickness of the pencil trace without which it would not exist".

An illuminating way of conceiving of the spirit and the god within, is to think of them as a duality in which the two elements come into and out of phase. That is, one can think of the latter as generous and always ready to join up with others, but which is kept on a tight rein by the wary, self-interested spirit until such times that it is to the organism's advantage to let go and proceed faster.

❖ 13. More examples of POF recognised or ignored.
1.Religion and the Spanish inquisition. During the centuries succeeding the crucifixion of Jesus his message of peace and love and the brotherhood of humankind, was subject to a process of analysis, interpretation, and logical development, much of which was coloured by the beliefs and judgements of the interpreters. An inflexible dogma was evolved which put on trial anyone who committed heresy by refusing to accept the guidance of the church. Suspected heretics were arrested, interrogated and tried; ***the use of torture was approved by Innocent IV*** in 1252. This was arranged by the leaders of a creed which advocated universal love, brotherhood and forgiveness. (POF violated – logic misapplied)

It is clear therefore that if a person is presented with a theory based simply on logic (mathematics is basically logic as shown by Whitehead and Russell), he will accept it as long as it doesn't stretch the god within too far as it adjusts to accommodate the process of reasoning in the conscious mind, which executes the logical manoevres of assimilating the theory.

2. In the fifteenth century William Tyndale was strangled and burned at the stake for translating the bible from Latin to English. Clearly the people responsible for these abominations were obstructing the spreading of the teachings of Jesus, and had lost sight of his original teachings which were intended to reduce the sufferings of humankind; they had

also lost contact with whatever feelings of compassion they themselves might have had, lead or seduced by a chain of logical analysis and interpretation, to distort his teachings so that they were inflicting the very suffering that they were meant to eradicate. This is clearly a case of the law of the primacy of feeling being violated by misused logic. (POF violated)

3. In western societies cannibalism is condemned as barbaric and the thought of eating human flesh is horrible. This is a feeling of revulsion. Logic might say, sensibly, "these are hard times, no food in the shops, it makes good sense to use the dead as a source of nutrition until things get better, and if one eats only people who've died a natural death nobody gets hurt. This makes good sense." Natural feelings of revulsion, however, would veto the idea.

Within a month or two, however, when conditions were at starvation level, people would be thinking seriously about the idea. In other words, the feelings of the starving people, which at first were utterly opposed to the idea of cannibalism, would have gradually given way to arrive at a state of balance with logic. The only imbalance permitted is for feeling to be stronger than logic and in this situation feeling, the feeling of starvation, could not fail to arrive at a state of balance with the logic so it is obviously a case of the law of POF being observed. Feeling and logic in agreement.

4. Everybody laughs and cries in the same language, the language of feeling and emotion. Feelings, especially emotions, are supra-national and universally recognized, anger, love, hate, sorrow, fear, surprise, sadness, hope etc. Everyone, no matter what his country or language, experiences and understands these feelings; and this even extends to animals, dogs, cats etc. who share some of our emotions. The language of words is the precise opposite of this (logos = Greek for word). It is based on and promotes differences. There are hundreds of languages each of which is understood by only a fraction of the world's population whereas the language of feeling is inborn and universally understood. This extends naturally to the arts, pictures, music and so on, where perception is immediate, lateral, intuitive, and where feelings – of pity, drama, horror,

peace, outrage, admiration, humour, interest, pleasure, or philosophical contemplation, are the object and result.

Logical/sequential perception relates more to science where the use of logic is fundamental, but both of these observations relate to the consumption of art and science; cutting-edge creativity in both the arts and sciences relies very heavily on feeling and intuition (god within), retaining only the necessary modicum of logic. In both science and art, in this example, the law of POF is observed.

5. The beheading of Ken Bigley in Iraq. Ken Bigley was beheaded while his family prayed for his release. His captors were willing to release him in exchange for the release by Tony Blair, of a couple of their own people held captive in the UK.. Blair, hidebound by inflexible logic and a short-sighted determination not to bargain with the captors, (since it could lead to more bargaining and be seen as a sign of weakness, which presumably he believed), ignored whatever feelings of compassion he might have had and allowed the appalling business to go ahead. If he had retained his better judgement and objectivity, listened to his feelings and those of Ken Bigley's family, and ignored the self -perpetuating logic, he would have released the prisoners, made many people happy and probably set in motion a continuing and ongoing relaxation of confrontational terrorism. This is clearly a case of logic misapplied and POF violated.

6. There is an account somewhere in the annals of scholastic philosophy, of the devout Christian Thomas Aquinas speculating with fellow-divines as to how many angels can dance on the head of a pin. This is a clear case of someone following a complex, seductive chain of deceiving logical deduction without checking for material evidence, and being led in a circle to a ridiculous and meaningless conclusion. Logic had him by the nose and led him in a circle as it so often does. (POF violated).

7. In an interview Beethoven advised a student composer who was seeking guidance, that if he should come to a difficult bit in a work, he should forget the rules of composition and write what felt right. There is a phrase in his 7th symphony, a quiet one which comes between two loud, assertive ones and tells us that he still believes in himself, he still has faith, but it is

very unlikely that he sat there thinking, in a detached, cerebral manner, "I will put in a phrase to express the fact that I still believe in myself." He would, of course, have been following a flow of feeling in which the self-believing phrase occurred naturally. (Confusing rules and logic discarded, POF observed).

8. Despite the dreadful suffering in Africa caused by sexually transmitted disease, the catholic church refuses to let its adherents use the condom, which is a barrier to the germs which cause the disease. If he could cast off the mass of contradictory dogma that has accumulated over the centuries as a result of logical analysis and interpretation, the Pope would take pity, as Jesus himself would, I believe, and immediately order an airlift of condoms to the stricken regions. (Erroneous logic wins, POF violated).

9. To continue this criticism of the actions of the Catholic Church, the logic employed by the Pope and his minions leads him even into advising against research into the origin of the universe 'because it is interfering with God's work and prying into the moment of creation', or words very much to that effect. Extending this argument would involve inveighing against the inspiring philosophical explorations of the ancient Greeks, which are an important part of humankind's cultural heritage. (POF violated)

10. The preceding items demonstrate very clearly that the inadvertent misuse of logic often leads by an insidious process to the perversion of an originally beneficent philosophy so that it becomes an instrument of oppression, e.g. communist Russia and China where preservation of the party machine was seen, correctly, to be vital to the continuing growth and survival of the movement, but criticism of which was interpreted as an attack upon its basic premises and philosophy; so much so that survival of the party came to be more important than the welfare of the proletariat for whom it was designed. The result was that millions were sent to die in Siberia, the echelons of power were subject to recurring purges, and a million intellectuals were liquidated. In China too, millions perished in murderous infighting between quarrelling factions in Mao's

unstable government, and on 'the Long March'. (Deceiving logic wins, POF violated).

11. In a romantic relationship the feeling, the 'chemistry' between the protagonists is of primary importance, but it is also essential that they have something in common, something to talk about. In such a relationship the feeling would be good but there might be very little to talk about which would make for long stretches of silence and boredom. It could improve as the couple found new activities together, or at worst would fizzle out within a few days or weeks. Conversely, if the feeling was bad, even though there was lots in common, the bad feeling would render the relationship unworkable and it would crash to the ground in flames. It must feel good. If now one had a relationship where the basic feeling was good and there was a lot in common, the chemistry, too, would be very good thus making for an excellent relationship. Such relationships all have two main components, the basic feeling of sexual attraction, and some common interest or activity about which to talk, (words), that is, feeling and logic. Balance between these last two is essential for a good relationship. (POF observed).

Feelings are vital and, generally speaking, should not be suppressed, i.e. feeling is prime. There is a mental illness that exhibits a deficiency in feeling. It is called 'anosognia' and the person suffering it while remaining perfectly aware and capable intellectually, has little if any emotional response to stimulae, so that life for him is a colourless, lukewarm affair. (Ref.5d)

12. A quotation from 'The History of Western Philosophy' by Bertrand Russell. 'Berkeley......For him there are only minds and their ideas; ...... but he still failed to grasp all the consequences of the ....principles that he took from Locke. If he had been completely consistent, he would have denied knowledge of God and of all minds except his own. From such denial he was held back by his ***feelings*** as a clergyman and as a social being (author's italics). (Fabricated, deceiving logic loses, POF observed).

13. A somewhat fanciful example. I am working at my PC and either it is 'playing up' a little or I am losing control, (neglecting the fact that it

could be both). I begin to get a bit hot under the collar so much so that within minutes, enraged, I am possessed by a desire to smash it to bits, (but I know that this would be wrong and knowledge is unassailable) and have lost all capability of carefully working out just what I am doing wrong – for the problem, as usual, is very unlikely to be with the computer, but caused by my own impatience. Here then we have logic (the PC) versus feeling *–my own impatient angry feelings* – and my only course of action is to accept the truth (highly probable) that the PC is right and I must wait for my feelings of impatience to subside. Twenty minutes later, having cooled down, I resume the exercise which either goes smoothly to a clear resolution, or starts throwing up problems again. If the latter is the case, I go slowly and carefully, keeping a firm hold on myself and my feelings until the problem is resolved and logic and feeling no longer contradict. In both of these examples it is clearly a case of logic versus feeling, and in both cases POF was observed.

14. A person is charged with committing a crime but the police were unable to prove his guilt in court so he is discharged not guilty. Unfortunately for the man, despite his protestations of innocence, the policemen who discovered the crime are uneasy, there is still a doubt in their minds, a 'hunch' (or feeling) that all is not yet clear and settled. The case is left open although technically it should be closed.

Six months later the detective in charge of the case is watching a film in the local cinema, when the events on screen trigger a memory connected with the case, and in a blinding flash he sees how the criminal did the job. The miscreant is re-arrested, tried and found guilty. Case closed. (POF observed. The officer could have closed the case but could not rid himself of a niggling little feeling of doubt which was created and registered, as in all the other examples, by a dialogue between the conscious mind and the god within).

15. I am in my new flat and have some pictures to put on a wall. One of them - my favourite, I put right in the centre of the wall. I look at the other pictures but can't decide which one to put up next. I could just be content with the one, my favourite, which would be perfectly OK but a certain inappropriate meanness or frugality makes me decide to keep

the lot. Finally I choose one of them since it has some of the qualities of my favourite. Again I am not sure which to put up next so I do the same thing again; I choose one that's similar to the last one or the one before. I keep on like this until they are all up there and step back to see the effect. It is disappointing. A general mass of insipid effects that doesn't amount to much of an experience, in the centre of which my beautifulfavourite brightly glows. This one was chosen by feeling and without an instant's hesitation. The others were chosen by a process of cold, unfeeling logic. (logic wins, POF violated).

16. It is 8.45am. and I'm about to leave my home to go to work, but as I get to the door of my room I get a slightfeeling of unease. I stand for a moment wondering. "Perhaps I've forgotten something?" I go through my bag, check my pockets, look on my list of things to be done. No luck. Everything seems OK.. I retrace my steps and notice through the window that it's raining. Wellingtons! I look at my feet and see that I am still wearing my slippers! I had forgotten to put on my trainers which I usually wear when going to work. The point I am making here is that before checking out all aspects of the procedure of going to work, one would logically examine each step to get to the conclusion that all was well, and yet the troubling feeling of unease, which was the work of the omniscient and all-seeing GW, the active element in the principle of POF at the seat of feeling, would refuse to go. (POF observed)

17. The mechanism in POF can be illustrated by comparing it to the experience of being introduced by a friend to a symphony by a composer with whose music one is totally unacquainted. While one's friend is going into paroxysms of delight one can hear only a cacophony of discordant scrapings, whistling and braying noises. One continues, however, to acquiesce to his exhortations and continues, over the next few days, to play the CD in the background as one works until suddenly, one evening, one finds oneself thinking, "I get it! I'm beginning to enjoy this – it's great!" One is beginning to 'get the feeling' and feelings, generally, don't lie. The confusing logic, structure and novel forms of the orchestration are resolved into a solid, satisfying and enjoyable feeling, and are shown in this case to contain truth and beauty. The point being made here is

that the feeling was vital in showing the value of the logic. If, of course, the music was bad or without value, one would play it until the cows came home all to no effect, and the bad feeling would insist that one turn the CD player off. (POF observed; the feeling was given sufficient time to adjust to the logic.)

It should be noted that in any scenario where POF applies, the relevant feeling may be any one of the feelings, or combination of feelings, enumerated at the beginning of this section (POF) i.e.physical, mental or emotional.

❖ 14. The phantom limb.
This phenomenon occurs when a given person has lost a limb, through accident or amputation etc.. It is the illusion felt by the amputee, that the limb is still in place, because he can feel it. One reason advanced for its existence is that the nerve endings are still sending signals to the brain. This would fit perfectly into TMM because according to TMM the electro-chemical pulses in the nerves and brain are turned into perceptions by the spirit – the seat of, among other things - perception. It is also possible that a tenuous limb actually *is* there, as in the Kirlians' leaf-lobe. When I think of, or imagine, a person in the afterlife, I think of the spirit of the person fleshed out and given atenuous materiality, invisible of course unless appearing as a ghost, by the PP which must accompany him/her, and I think of it as human in form, and then I remember the words which came to me in my ultra-idealistic lonely days, namely – "the spirit informs the flesh". What this means is that any given face, in the present life, (especially in younger, not yet fully-formed natures) if in a state of repose, would appear smooth and at peace. At other times however, in animated conversation say, it would display the subject's current mood, i.e. loving, hating, joyful, sad, hopeful, wistful and so on, and these facial variations are not simply the result of muscular activity; the actual flesh of the face is also animated by the spirit, as I have observed many times and carefully noted. These observations also apply to 'the Elixir of Life' as discussed later in the essay, i.e. a youthful spirit in an older person can transform the appearance from older to considerably younger.

Returning to the phantom limb, and the possibility that it continues to exist, as may the Kirlians' leaf-lobe, the limb would remain in place – still be there but in PP form only. It would be as though this part of the body, which has died and is, maybe, decaying in some foreign field, is still there in spirit and is effectively already in the afterlife, but still attached to the spiritual component of the living body.

It is interesting, as an aside, to consider the fact that feelings do take time to change, whereas the speed of change of logic is limited only by the physical characteristics of the system employed, e.g. computers and 'fast logic' (a technical term employed by information technology, where-in an event may endure for less than a billionth of a second).

CHAPTER 3

Godel's theorem and to feel is to know

1. Head, heart and gut; 2. Memory; 3. Kurt Gödel and the Incompleteness Theorem; 4. Intuition and instinct; 5. Expressions describing intuitively perceived truths; 6. To feel is to know (TFITK). 7. Examples of TFITK..

❖ 1. Head, heart and gut.
The head, as everyone knows, is responsible for calculation, analysis, classifying, general intellectual functions, judgement and attention directing. According to TMM however, it also separates material that is useful only over the short time span (short-term memory), from that which it is necessary to retain on a long-term, maybe permanent, basis. The short-term material remains at head-level and gradually fades over a period of minutes or hours. The second lot of material is useful over the long-term so it is directed by the head and brain into various loci in the body, neck, shoulders, armpits, heart and various places in the chest and torso, the lowest level found by me so far being a couple of inches above the reproductive organs in the abdomen. It seems likely that the oldest memories are found in the lower parts of the body. These sites are pretty certain but it is possible that memories are lodged throughout the body in muscles and possibly bones. There is even a case in which it is possible that memories exist outside the head and upper body in the PP that surrounds them. (Since composing it, the last sentence has been verified).

The heart is the home of the spirit or essential self and it is the place where voluntary mental activity, imagination, thoughts, words and images originate and combine with material from the unconscious, before rising to join activity in the head. By 'essential self' I mean the spirit inhabiting and animating that particular body, as opposed to the personality or mind as a whole which includes the head. The spirit, which used to be called, with reason, 'the holy ghost', is the essence of the person, and it departs, at death, for the afterlife. I think it is very probable that it is accompanied, in doing so, by the rest of the mind and personality, including the god within. The heart and spirit are also the origin of courage, drive, willpower and initiative, all 'intentional'characteristics to quote William James.

A very good argument for the existence of the spirit or soul, is the question of freewill. I think it would be generally accepted that no computer, no matter how large and complex, not even 'neural networks' based or those running heuristic programs exhibits the slightest suspicion of freewill. All computers, both digital and analogue, are essentially logic-based and rely for their operations and general behaviour, on counting and measuring, (digital), or model-making, (analogue), and some kind of human programming. They are not autonomous. A robot too would rely on information derived from its environment to determine its actions and this too would be largely translated into digital form, that is, states of its behaviour expressed in terms of logical operations in the form of '1' or '0'. This information might then be run through 'algorithms', that is, formal procedures to arrive at a result or instruction which could then be used to actuate a limb, speaker or some other function.

It is clear, then, that not even the largest of our computers displays freewill, even though the density of packing on a 'chip' as seen in current integrated circuits and microprocessors is such that the separation between elements of the device is measured in molecules. In other words the organisational density of our computers is as great as that of brain tissue, including the brain of a common house-fly, which, of course, does have freewill although this assertion may be challenged by some people. I see it though as axiomatic that any form of life is intrinsically possessed of a measure of free will. Freewill itself could be defined by the relevant lifeform's capacity to operate in a non-random manner, and unsupervised by a superior intelligence.

One is now prompted to ask "if our largest computer is millions of times more complex than the cubic millimetre brain of a house-fly, how is it that the fly has freewill and the computer doesn't?" The answer is that the fly has a metaphysical component, the spirit, and the computer doesn't. (One must assume – a slightly comical idea but true I think - that the fly has, in addition to its spirit, a tiny god within, since it possesses an abdomen and a thorax).

In the period of intense loneliness and isolation to which I have referred elsewhere, I remember odd bits of information coming to me, e.g. 'all gods are good but some serve evil ends', and 'the spirit informs the flesh'. The former I interpret as 'the god within is essentially good, but being subject in the last analysis to the spirit in the heart, may be obliged to supply the energy for evil or immoral actions, as directed by the head, which is the final arbiter of the individual's activities.' I used these messages, plus other information from diverse sources, as part of a basis for my theory. What all this means, however, put in everyday human language, is that one lives one's life consciously, choosing this course of action over that, making decisions, planning, doing physical work and/or mental work and the physical energy necessary for these activities is supplied by food in the gut being broken down and turned into glucose to fuel the muscles, repair organs and so on. Metaphysical energy can be broken down into four distinct kinds. These are listed in item 85 in the addenda.

The gut is the home of the GW, with its appetites, sexual urges, humour, creativity, enthusiasm and, through its status as a divinity connected with almost seven billion other gods within, (the population of the earth), its capacity to transcend space and time to get access to distant regions of being and existence. The gut is also the home of the unconscious mind, of which the god within is the engine.

❖ 2. Memory

Memory (fig. 6) and other mental faculties are usually assumed to be confined to the head, although Aristotle believed, correctly I think, that thoughts arise in the heart, especially purposeful ones, though they can arise anywhere in the body I think. The fact is, as already stated, that memories are stored throughout the head, chest, shoulders and abdomen (Ref. 15). Muscles contain memories. Very old memories are located in the abdomen, more recent ones in the heart, chest and shoulders and very recent short-term ones in the head but not necessarily in the brain alone. These facts were obtained partly by subjective experiences and partly from information collected from various diverse sources. At least one of them was an account of a heart transplant which produced in the recipient, personality traits which were not there before, but were known to exist in the donor. This raises, in passing, the horrible possibilities of animal/human transplants producing similar effects. One can recall memories from one's own unconscious (GW), but it is a fact, I think, that one also has access, at times, to the unconscious minds of others. This is because it is the nature of the god within, the motor of the unconscious, to reach out to connect with other gods in other people, so much so that an individual god could have access, as written earlier, to the contents of every unconscious mind on the planet. It is also the nature, one would surmise, of a metaphysical entity of the status of a divinity, to transcend space and time.

These statements are supported by recent research which revealed neural feedback circuits associated with memory which were found in the heart. This information was obtained from a TV program (CH.4) on 29/06/09. The expression 'learning by heart' is an intuitively perceived description of what happens when one engages in learning by rote. In this activity the memories are actually stored in and around the heart, lungs and ribcage. General knowledge, lessons learned by rote, personalities and attributes of one's friends, social contacts, and most if not all middle-term experience is stored here I think.

There was a TV program a few days ago in which a scientist revealed, with the aid of various electronic instruments, that when a person makes a so-called conscious decision, the decision has already long-since been made somewhere in the body or unconscious mind, and the scientist can tell what the subject is going to decide, as long as six seconds before he does decide! I think my theory can account for this extra-ordinary fact. What must happen is that there is a fluctuating psychoplasmatic complex containing feelings, images and words, at the level of the heart or spirit while the subject is making up his mind. This complex will be influenced by the presence of the GW (unconscious). At some point in this process the activities of the GW will achieve balance, or accord, with those of the heart, and a dominant feeling will emerge which slowly makes its way to head-level where it comes into consciousness and is expressed as the best choice to make, so the subject makes the decision.

It is also a fact, I think, as already written, that memories, and psychoplasmatic activity, abound in the psychoplasma surrounding the head and upper body, as explained in Chapter 6 (item 5). As a result of that minor exposition I was lead to speculate on the significance of the particular region of psychoplasmatic space from which the handful of PP was grabbed. I decided that it was possible that that particular region was where Kenny Everett stored his nonsense, (and I mine, since human physiological similarities probably had a say in the matter), and that it was possible that the space around the head was 'compartmentalised' like a filing system. On this basis clear, rational thought would be conducted in the space in front of the forehead or cerebrum and also within the cortex itself.

Bearing all this in mind I decided to try relaxing while sitting at my PC, placing my awareness in my feet and awaiting developments. I became aware i.e. my attention seemed to focus, or become focused, at a point two feet above and six inches behind the crown of my head, oddly enough at the 'nonsense point'. It felt as if I was being observed - by myself I think – self-observation, but only as an object, not in terms of my actions.

❖ 3. Kurt Gödel and the Incompleteness Theorem.
Gödel was a mathematician who, in 1931, formulated the above theorem which stated, expressed simply, that it is impossible to formally prove, within a given system, any mathematical proposition without going outside the system, which meant that some propositions could be seen to be true but could not be proved.

This imposed a fundamental limitation on the usefulness of logic and mathematics – mathematics is basically logic - so that ultimately we must rely on intuition (feeling), which is more primary and basic than either logic or mathematics, and is the matrix of all creative endeavor. Einstein too put a great emphasis on intuition and believed that a viable theory must be based in it. This means, in ordinary language, that the mathematician may be aware intuitively, i.e. can clearly see, that such and such a mathematical fact is true, but has no formal or symbolic way of putting this fact down on paper, or demonstrating its truth without encountering a contradiction. This nicely illustrates my difficulty in providing a proof of the existence of the afterlife, which ***I know without any shadow of doubt*** to be there. I have no way, at present, of giving a verbal proof; (but see addenda 115); the only way to surmount this problem is to help the seeker after truth to **experience** its reality by gaining telepathic access to it, in the same way that I have experienced it. That is, proof can be gained by the mechanism of TFITK (see chap.3. item 6)

An interesting and cheering aspect of the existence of the afterlife, digressing for a moment, in view of the fact that it is possible to communicate with people therein who died two or three millennia ago, is that it makes it practically certain that one may join one's departed friends and relatives there, since they, according to this fact, will almost certainly be living there for at least two or three millennia. The 'downside' is that one may also bump into the man who's wife one stole, or be punished for other crimes. There is no doubt that we shall all be going there so wrong-doers should beware. This passage on the incompleteness theorem is pertinent to the theory since it calls the certainty of logic and mathematics into question, thereby asserting the more fundamental nature of feeling out of which grows intuition, the mother of all creativity.

❖ 4. Intuition and instinct.

Intuition can be classified, with instinct, as being a variant of, or related to, feeling, which is based in the gut and is a function of the god within. Although it is related to feeling, however, it involves a large degree of intellect and therefore is a function, I think, of the conscious mind and the unconscious one working together. Feeling is located, basically, in the personal unconscious but one's intuitive faculty serves a dual function, the first of which is the direct acquisition of knowledge without the use of reason. The second concerns creativity which is - as is generally believed - a major function of the unconscious. Intuition works through, and by virtue of, the fact that the GW is the active, guiding force of the the personal unconscious, but in addition to this, due to the admixture of BP that mixes with spirit (S) its operations are present in the background of the conscious mind, in co-operation with which it displays the intuitive faculty. As a consequence of its function as an unknowable quantity or entity, the internal operations of which are inaccessible to scientific curiosity, it is not possible to look into the heart of the creative act, an act which results in a hitherto unknown, nonexistent entity be it a painting, symphony or scientific theory. All of these things come to us via the operations of the god within and its mysterious and forever hidden activities. This, essentially, is the content and conclusion of, and the reason for, the preceding section ' Kurt Godel and the the incompleteness theorem' where it is seen that an arithmetical (or mathematical) theorem cannot be formally proved by use of mathematical symbols, although its originator may know that it is true. It is doomed to remain an intuitively perceived though inexpressible fact. . Instinct, I think, relates more to natural, immediate action, that is, automatic unlearned response to new, previously unknown situations. The superiority of intuition over logic in matters of creativity is another illustration of The Primacy of Feeling.

Feeling is more important and fundamental than any other human faculty, and it is continuous rather than linked, unlike logic which is the latter. It is a rope and as such can be stetched whereas logic is a chain.

❖ 5. Expressions describing intuitively perceived truths.
There are many verbal expressions which occur in day to day conversation and the words of popular songs which express feelings or situations in a direct, immediate manner and which are not the result of considered cogitation but spring immediately into the mind and thence into the current demotic. Many of them express intuitively, and practically unconsciously, perceived facts relating to psi and the TMM. I can quote a few here but I'm sure many more exist:

- To 'give someone the cold shoulder' = the retention of warm, friendly PP instead of letting it flow.
- Casting round for ideas = mentally trawling the surrounding PP for ideas or information.
- Casting the mind back = delving into the PP behind oneself at and above head-level to remember what one did yesterday or in the days and years before.

- Sixties hippy 'getting it together' = the act of drawing one's mind (PP) together and concentrating it.
- Feeling 'cold and lonely' = shortage or tenuity of PP which is largely derived from and shared with companions.
- Another song: 'Got to get that man right out of my hair' = get man's sticky PP out of hair.
- Hard work 'takes it out of you' = exhausts one's PP reserves as well as physical energy.
- 'He hasn't got it in him' = PP (or S +BP) which can be equated with sexual energy.
- Another song: 'You've got me on the end of a string, please don't let me drop' = One lover attached to another by PP, and pleading for mercy.
- A person is described as 'a wet blanket'. His mind (PP), as well as his actions, will mix with and pull down that of his associates.

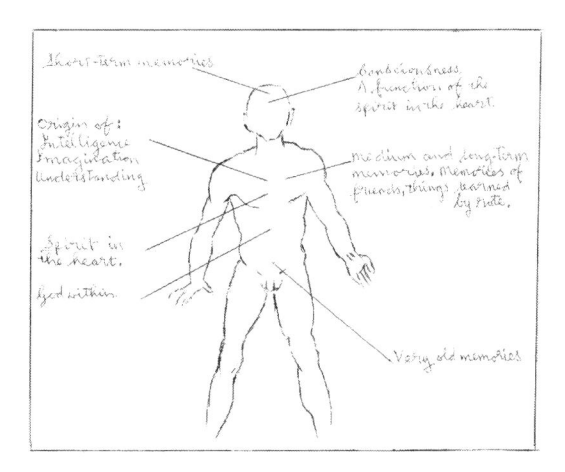

Fig. 6. Distribution of memories in the human body. .

- Feeling heavy-hearted = ratio in the heart of BP to spiritual substance is larger than optimal value
- Feeling light-hearted = " " equal to"
- When one feels 'unsettled' = PP distribution is fluctuating and erratic.
- 'He's not all there' = slow, erratic person with faulty perceptions and erroneous thinking processes.

I've heard this expression used in my home town, and what it really means is, that the spirit of the person in question has become partially dissociated from the god within, so that the person's conscious mind or psychoplasma and unconscious mind or bioplasma are stretched apart and are centred in different places.

These expressions are a useful aid as illustrations or confirmations of models or hypotheses (thought experiments) which facilitate understanding the mechanisms of psi and consolidating the knowledge thereby so far obtained, and their utility and accuracy reside in the fact that they are results of the spirit, immediate and spontaneous, not considered, working with, and being informed by, the personal unconscious motivated by the GW which never lies.

❖ 6. To Feel is to Know. (TFITK)

Feeling is more important and fundamental than any other human faculty. Feelings between people (and pets etc.) are real; they are conveyed via psychoplasma (PP) which is a substance possessing mass. They also exist within the overall substantial reality which is the Bioplasmatic Envelope which unites all life on Earth, including the afterlife. In a conversation one is aware of sounds, words, facial and bodily movements of one's interlocutor, which may be simply an awareness of his/her presence plus a feeling of engagement. In some cases the feeling will be very strong; in others it may be hardly perceptible and it may take many forms. The feeling of friendship is a case in point. It is a warm feeling and if one is separated unexpectedly for some reason, from a close friend one feels the cold seeping into one's middle where the GW resides. There is a moment burnt into my memory of myself aged twenty and my two closest friends, having been together all day, having to say goodbye for the day to depart for our respective homes. There was a feeling of warmth uniting us and as we reluctantly said our farewells, I could feel a cold emptiness invading me, and I knew that we all felt it, although I also knew that when I got home it would disappear because they would be there tomorrow, and my family too were, of course, a source of warmth and security. The feeling of warmth was a psychoplasmatic attachment between the three of us, which would thin out as we parted so that it would become subliminal, but it would remain there as long as we remained friends, that is, as we talked with our parents and siblings we would forget our friends although we would remain subliminally attached to them. This attachment would no longer be a purely psychoplasmatic one, which would be conscious, but would tend to acquire a greater proportion of bioplasma operating then between our unconscious minds. We would also, of course, be returning to longstanding psychoplasmatic attachments with members of our families.

In cases where the main feeling is not one of friendship, the feeling will vary according to the nature of the relationship. A conversation between a tutor and a student about the relevant subject will generate a cool, steady feeling; the student may be sitting attentively near his tutor aware of his physical presence, and there would be a light but friendly attachment and a feeling in the student of respect for his tutor. An argument between a man and his wife will generate a feeling, a PP attachment, which may vary intermittently and jerkily between thick, heavy, strong, viscous and thin, weak and tenuous. This is how it will feel because the feeling between two people is a function of the nature of the PP attachment. PP is real. It is a substance; although the feeling of attachment may be extremely light and almost non-existent, but in all cases it will be real.

Sometimes one encounters situations in life, which involve talk, argument, discussion or persuasion with or by somebody where it is difficult to determine the truth. As he talks one's thoughts, reasoning and imagining, one's intellectual operations, are confusing and inconclusive. Often, however, in this kind of situation one's thoughts may coalesce into a feeling vis-à-vis this somebody which gives an absolutely indisputable guide to the truth, and may even, sometimes, reveal one's interlocutor as a joker who is leading one up the garden path! The feeling will not let you down, it is real and is conveyed by PP – a real, tangible substance. I have given the title 'To Feel Is To Know'(TFITK) to this mechanism. (See addenda 107).

❖ There is a popular belief that dogs can smell fear and when they do they will attack. This is probably true because a frightened person will give off odours which are detectable by the extremely sensitive canine nose. It would seem quite possible, however, that in addition to this mechanism what is actually happening here is that the dog may be nervous, uncertain and consequently liable to attack. The victim is also nervous and may fear that he may be bitten, so that the feeling between the two (the PP or telepathic connection) is highly volatile. In this situation the dog, if it senses fear or uncertainty in the person and, being driven by instinct rather than reason, will naturally attack. A similar situation sometimes occurs when a lone pedestrian is waylaid by loutish young yobbos. If he keeps calm (keeps control of his feelings – that is, the ambient PP, –which connects with that of the yobbos), his calmness and assurance will be felt and he can often defuse the situation, but if he loses it he will incur their derision and they may attack.

Part of the significance of TFITK is that the mechanism, (the feeling itself) in all cases, though often accompanied by related body language, reveals the truth implicit in any social situation. It reveals or betrays the thoughts, feelings or state of mind of the other person or people concerned.

❖ 7. Examples of To Feel is to Know (TFITK)

1.We know intuitively that animals are not aware of the fact that that someday they will die. We feel this and therefore know it.

3. One enters a room and is instantly struck by the prevailing atmosphere which may be a feeling of tension,gaiety, solemnity, quiet industry, piety, anxiety, a buzz of excitement, peace etc. One's perception of the atmosphere may include visual and/or auditory clues but the perception is usually so immediate that it must be put down to a direct apprehension of the feeling or atmosphere (PP) in the room. (you could cut the atmosphere with a knife).

4. The phenomenon of making a distant person turn round by staring at him from behind is also an instance of TFITK. Its mechanism is described in the main text.

5. I walked from the garden into the dining room. The serving hatch was closed and from behind it I heard thesound of two voices, both women.

I had made no sound on entering but I suddenly realised that one of the ladies was becoming aware of my presence and for some unknown reason was getting angry with me. As I stood listening the feeling got stronger and stronger and the lady was getting more and more angry so I tactfully removed myself to another room. I think here again it was a clear case of interaction between her mind and my own via the intervening PP. She could feel my presence and consequently had no doubt that I was there.

6. One is out and about and suddenly catches sight of a person who is not quite a friend but nevertheless, a pleasant social contact. There is no eye contact but as he walks off in another direction one is suddenly quite sure that this is not because he didn't notice one's presence, but is a case of being 'cut dead' or snubbed. There is a definite, cool, absent feeling, a gap in continuity of feeling, which betrays his apparent failure to see one's presence. On the other hand, if he really hadn't seen you it would feel totally different, warm and friendly and you would shout his name and call out "hello".

7. There is a certain person who every now and then knocks at one's door to ask if he can borrow two or three pounds. Normally one does not lend money but one has weakened in this case and wants to put things straightby refusing. He turns up again one day and normally one would simply say, "I'm sorry but I've decided not to lend anymore money to anybody" and he would go away. On this occasion, however, losing one's resolution one 'fudges it' and lies instead saying "I'm sorry I don't have any change" but one says it in such a way that he knows instantly that one is lying, not just by the hesitant, uncertain body language which could originate from something else, but again by an unmistakable feeling i.e. close-quarters telepathy consequent on the strong PP connection which would automatically exist and which would betray one's insincerity, and he goes away surprised at one's uncharacteristic behaviour. Also, because one was oneself acutely aware of the feeling one knows that he was aware of one's lie.

8. One is sitting at table in the dining room, where the conversation is being monopolized by a certain loud, aggressive person. She suddenly

announces that she has to go upstairs to finish a task and in one's mind one cannot help but think, "thank God for that!" Unfortunately, for her, she senses one's thoughts via the change of feeling (PP) between her and oneself and goes away appropriately downcast.

CHAPTER 4

The universal unconscious and the afterlife

1. Telepathy; 2. Clairevoyance or second sight 3. Psychokinesis; 4. Unconscious mind suppression phenomena (UMSP) (poltergeists); 5. The kinetic energy used in UMSP; 6. The BPE and the discovery of the afterlife; 7. The BPE connects all life that exists or ever did exist; 8. The universal unconscious; 9. Levitation; 10. Out of the body experiences and consciousness; 11. Teleportation; 12. Dowsing and divination;

❖ **1. Telepathy**

The word telepathy means 'feeling from afar' and was coined by Frederick Myers of the Society for Psychic Research in Kensington, London at about the same time as the Society was founded in 1882. I think there may be two kinds of telepathy one of which would be called, more accurately, telecognition, since it would be a capacity for the transmission of descriptive and quantitative information. Normal 'common or garden' telepathy relies on PP but is perceived as a feeling which may or may not contain discrete items of information such as words. Telecognition, low-level experiences of which I have had myself, also relies for its working on PP plus a kind of divination that I shall write about in my next essay.

In telepathy the spiritual component in PP is the medium, except for unconscious telepathy where I think BP is the medium, and the transmission of information is effected by means of a psychoplasmatic connection, via the CC, of one mind to another; this may happen directly mind to mind if the two are near each other, in which case it would be more appropriately known as the principle of 'To Feel Is to Know' (TFITK) - see chap. 3 item or via the collective conscious if they are far apart. Telepathy in normal life as opposed to experimental situations, is concerned with the conveyance of feelings and emotions rather than words, cold facts, prices of goods etc., although the PP may contain words and images, and/or any feelings may be associated with specific words or images. If one is at a party for instance, one will be sensing the closeness, distance, or behaviour of other people and oneself via the fluctuating PP connections, that is - feelings- as well as direct visual, auditory, olfactory and tactile channels. One will feel the atmosphere which is merely PP in animated motion. When a complex telepathic communication occurs, such as the one that happened between a nurse and myself, some years ago during a sojourn in hospital, what happened was that she knew that I was looking for a girlfriend and guessed when I said, "hello", that I was about to make advances. She moved away but I was strongly attached to her telepathically and got a changing mixture of words and feelings from her. I was strongly conscious that she was thinking, "some hope you've got mate! You're not in my league. Get lost". I doubt that she was thinking these exact words, she may not have been thinking in words at all, but there is no doubt at all that she was experiencing corresponding feelings. Her subjective state would have been a flow of feelings such as, scornful, mocking, superior, amused, bored, dismissive and the corresponding words as I perceived them, if I did perceive any words rather than just feelings, would have been inferred from them, by me via my faculties of interpretation. This is an example of 'to feel is to know' as, in the last analysis, is any telepathic communication. This kind of experience would be quite commonplace and frequently encountered.

 I, personally, seem to have acquired the power to establish telepathic contact simply by thinking of someone, a friend, relative, or a departed parent, or public figure in the present life or the afterlife, that is, anyone, anywhere, anytime, and when it happens that person is aware of me as a presence in his /her mind; that is, he/she is simply thinking of me without being aware of the fact that he is being made to do so by the fact that I am thinking of him. It often happens by accident. I can be reading something when a word or event reminds me of someone and I am immediately in touch with them. It can be very irritating. I find it almost impossible to think about someone in a detached, objective manner, without finding myself, in seconds, in telepathic contact even with perfect strangers.

The knowledge that I've acquired during my last few years of thinking and observing, elevates my formerly humble self to quite astonishing heights of objectivity and gives me extra-ordinary powers for good or evil as I see fit. Naturally I never use my powers for the latter in fact I never deliberately use them at all – I try to live a normal, natural life as far as possible. My knowledge has also given me some very weird experiences, which I shall describe later in the essay.

I find also that I can enter a building such as the many-floored and many-roomed warehouse that I share with about thirty other artists, and am immediately aware of the presence of two or three other friendly artists: I know that they are in the building because I can feel them, although I can't see them. I have checked this a couple of times and found it to be true. I should have checked every time but was prevented by circumstances from doing so. We are all connected via the medium of the collective conscious which is composed of psychoplasma. It is also certain that telepathy takes place during dreams and can sometimes lead to a spontaneous teleportation (see chap.8 item 6 bullet 9 and ref. 1c).

The 'material mind' is an umbrella term for the personal unconscious and conscious minds of an individual, which are made of bioplasma and psychoplasma respectively, and can grow to cover an area of one or two square kilometres or more. Higher animals too, apes, dogs, cats, horses also have a place in this system, as indeed, in various ways does all life on the planet.

❖ 2. Clairevoyance or second sight

Clairevoyance or second sight is the ability possessed by some people to see a distant scene or object by purely mental or spiritual means. The most well-known exponent of this ability in the 20th century was an Irish woman, Eileen Garrett, who founded the Parapsychological Foundation of New York. I would like to quote, word for word, an account of some of the remarkable things she could do, but the instant I begin to copy from text I feel and know that I am being robbed of my undoubted knowledge by certain rivals through telepathic means. In view of this I am forced to quote from my memory, which is not infallible. One of the more amazing things done by Mrs Garrett was to identify several books placed in envelopes without her knowledge of them, or that of the people who put them there, and then transferred by other people into different envelopes, and name the people, all unknown to her, to whom they belonged. This whole operation was rendered more impressive by the fact that the books were 2,000 miles away in her doctor's office as she performed the feat.

This is easily explained by the TMM which would be permeating both books, envelopes and office and would be connected, even if PP density dropped to its basal value, to the mind of Mrs Garrett, at all times, through PP on a conscious level and BP on an unconscious one. It is, of course, the conscious connection that astonishes since the unconscious one, though present, would be unperceived, but ultimately, over large distances conscious and unconscious would merge leaving pure basal or near-basal density PP.

❖ 3. Psychokinesis (PK)

PK is the ability possessed by some people to move material objects without touching them or using any contrivance or device to do so. I myself find that I have developed a modicum of such powers, although in my case, as in experimental situations generally, they are very capricious; they come and go. (this does not apply to my ability to connect telepathically with others, unless, oddly enough, I am trying to connect as part of an experiment, which itself is explained by my analysis of the reasons for the universally encountered nonreplicability of psi (see chap. 2 item 2).

I first attempted to investigate the mechanism of PK by assuming that it must be moved by a material attachment, i.e. bioplasma - BP - but since BP is a substance associated wholly with the GW and therefore the unconscious, this seemed insufficient and I immediately suspected that, since the spirit is the conscious animating and initiating component in the conscious mind, it must combine with bio-plasma to produce what I call psychoplasma, in which form it could permeate an object, get a grip on it and make it move. Some years ago I did some simple dice throwing experiments and found that when I threw and got a six I actually felt a slight elasticity between the palm of my hand and the die as it span to the ground. (At this point in time I see that to throw from the hand instead of a container could invalidate the experiment – but this does not alter the fact that the elastic attachment was felt). In confirmation of all this, I can say that I am certain that it is the materiality of BP which enables it to cause movement when it suffuses material objects but the movement is not, in my case at least, the result of a conscious 'willing' the object to move, but simply the result of 'expectant observation'. The actual movement is effected by the operation of the god within judiciously deploying BP. That is, on the many trials of PK which I have made, most of which were unsuccessful due to the 'nonreplicability' effect, those that did work happened as I sat or stood and simply watched the object (an umbrella PK demonstrator described in chapter 8 item 9)

I think, therefore, that the role of the spiritual substance (spirit) in psychoplasma, is as an agent by which the spirit gets a grip on the object which was my first assumption, and is an extension of the conscious mind which is the part that does the observing and directs the energy of the GW to move the object in question, rather than supplying energy itself. Indeed BP can and does work alone, notably as an agent of PK in poltergeist phenomena. (described in Chapter 4 item 4).

In formal experiments by J.B. Rhine it has been found that interest and motivation play an important part in getting positive results. This could be explained by the probability that high motivation would increase excitement and therefore enthusiasm (GW) resulting in a greater density of BP + S = PP. This is a very cogent explanation of the phenomenon of 'winning streaks' in gambling. The gambler has a win, gets excited, and another, gets even more excited and enthusiastic, goes on winning etc. until he hits a loss; then it is disappointment, negation, winding down to losing and normality.(Unless he has the sense to stop at once and pocket the cash!)

❖ 4. Unconscious Mind Suppression Phenomena. (UMSP) (poltergeist phenomena).

These phenomena usually occur in households where one or more pubescent or adolescent children are living. They are caused by tensions existing between the parents and children, who are in the process of casting off repressing aspects of their parents' value system, and establishing new identities with friends outside the family circle, thereby finding themselves in conflict with their parents. They will be reluctant to hurt or offend the parents and will force themselves, when at home, to stay in the mental strait jackets unknowingly imposed by them, so the superfluous energy will be 'driven underground'. That is, the conforming conscious mind will suppress the unconscious mind (the god within) which, unable to frezly influence the conscious mind, and therefore the subject's behaviour, which would happen naturally with new friends outside the family circle, will then divert the trapped energy and start playing tricks. Since the god's nature is essentially good, humorous, sometimes a little mischievous, and it is incapable of doing wrong or evil the tricks are usually harmless and amusing although this is not always so (Ref.11a). The cases of assault quoted in the reference must be the result of something, evil spirits or the spirits of evil people, getting into the heart of the person afflicted and influencing the person's actions. (I see nothing invalid in this idea. My belief in the spirit is absolute, so if there are good spirits, which there are, why should there not be evil ones, or disillusioned, vengeful ones?) The medium by which these pranks are made possible is BP (unconscious mind) whereas in cases of PK the medium is PP (conscious mind), although in both cases BP, as a substance possessing mass (I think), provides the necessary physical force acting on the object. The difference between the two is that PP contains 'spirit' which I think of as a presence or power rather than a substance (though I may be wrong here) and which can be consciously applied to produce physical effects, whereas BP alone works quite unconsciously. I read recently that Hereward Carrington, the well-known psychic researcher, did some experiments with a helper, designed to measure the weight, if any, of a rat's spirit. They did this by weighing a rat before and after its death, and found a difference of about 2.25 ounces, but it should be noted that similar experiments have been done with human beings, which gave ambiguous results. Also the weight he obtained would be influenced by the proximity of nearby living organisms. If Carrington's result was true, it could be explained as a result of the loss of the spirit *plus* the mind (PP) which would have departed to accompany the spirit into the AL.

Much UMSP appears to be very knowingly executed (doors opening and closing, lights turning on and off etc). This can be explained by the fact that the god within, the engine of the unconscious, usually goes about its business in a purposive and volitional way, knowing exactly what it is doing. In a complex situation where there are several people involved, the unconscious minds, which consist largely of BP deployed and actuated by the mischievous gods within may pool their energy thus acquiring enough at times, depending on circumstances, to move large, heavy items such as articles of furniture. The loud reports heard by Freud and Jung when they met in Vienna in 1909 and which Jung described as 'catalytic exteriorisation phenomena' (ref. 1d) were essentially of the same nature, one mind in conflict with another, forced by convention to suppress its true feelings. That is, Jung could not express his true feelings (which originate in the unconscious mind) on getting a shallow dismissive response to some remark that he made so the energy was directed by the god within (Jung describes a strong sensation of heat at the solar plexus - the seat of the god within) into the environment. I myself have had many experiences of recurring blowing of light-bulbs and central heating malfunctions which I see clearly now were the results of a period of mental illness which I was then suffering, and which caused me to suppress my real feelings and desires in submission to an impossible ideal.

This is a convenient and graphic illustration of the fact that these phenomena can happen even where there are no children present as my own experiences have dramatically demonstrated. The thing that is common to all three instances is that in both my own experiences, Jung's and the naturally occurring ones where children are present, there is a suppression of the activities of the personal unconscious of one or more of the people involved, due to restrictions placed by the conscious mind on the behaviour of the individual concerned, the protagonist, who is being forced unwillingly into uncomfortable and distressing situations.

An additional illustration of the mischievous nature of the god within and the fact that it alone, in conjunction with the spirit, is the ultimate source of paranormal phenomena can be found later in the essay. (Ref. 12).

❖ 5. Kinetic Energy expended in UMSP.
A given individual may have slept well, had plenty of rest, and sufficient food and drink to get him comfortably through his morning's work, but still feel tired and lethargic until he arrives at work, to become immersed in a pool of unconscious BP and conscious PP generated by

his workmates, with which, if positive, he flows together to untangle and invigorate himself so that he can do his job.

The minds, both conscious and unconscious, and the body are, of course very closely related, and are influenced for better or worse, by their psycho-plasmatic connections with others; they will be full of energy when the influence is positive and lethargic when negative. These connections provide the mental energy necessary for the body's motion. By mental energy I mean something different from ordinary physical energy which is derived from food in the stomach and intestines to supply chemical energy to make the muscles move, and which can be measured in joules or calories. I have, actually, in another part of the essay, (item 85 in addenda) already given names to this non-physical energy, described more exactly as 'metaphysical energy'. That is, energy supplied by the god within is 'sexual energy'(which could be thought of as enthusiasm) and that supplied by the spirit, which when it is mixed or compounded with BP thus making PP has two aspects, as (a) the origin or matrix, of much paranormal phenomena, mediumistic capacities for instance, and (b) as pure spiritual energy, in the heart, where it is associated with willpower, strength of mind, courage and the virtues, purpose, and intentionality in general.

In the lonely period mentioned earlier, one of the phrases which came into my mind was 'spirits ride on spirits'. What this meant was that each one of us carries around, in his/her mind, at various levels of consciousness, pictures or models of his friends and contacts and at the same time will be in fluctuating conscious and unconscious telepathic contact with them. Thus a host of mutually helpful relationships will exist. I used to think that the message, 'spirits ride on spirits' like other enigmatic statements, such as, 'all gods are good though some serve evilends', came from the unconscious, the GW, which may be true, but I now see that they were, sometimes, the spirit,that is, I myself perceiving a truth, and other times simply an observation that came from I know not where. The GW, incidentally, never tells a lie and never 'gets ill', as can the conscious mind, largely, I think, because though it is essentially good it is also amoral, and therefore not bound by rules, regulations and taboos as is the conscious mind.

When I was in hospital with a mental illness I noticed two individuals who seemed to find it impossible to get out of their chair or rise from the settee without a massive effort, or help from a nurse, and I myself sometimes had to endure this indignity. I think that what was really happening was that there was a conflict between feeling (GW) and knowing (head and heart) The patient, for various reasons, believes that it is impossible to combat the feeling of heaviness by rising from his seat. He believes it to be impossible. His memories, in psychoplasmatic form, congregate around his head to confirm this belief. I remember an occasion when I myself was in hospital with a mental illness when, at mealtimes, I used to find it impossible to sit down to my meals and had to eat while standing. One day, while conversing with a support worker there, I innocently sat down to my lunch, but, to my horror, a moment later I 'noticed' that I was sitting and instantly felt obliged to rise again! I had forgotten the disturbing thoughts which had been hanging round my head stopping me from sitting, and having noticed this fact, they all came back. I later escaped from this predicament by working assiduously at my art work, poetry, psi research and keeping a journal. It took a few days but eventually I did forget the disturbing thoughts and managed to sit down at meal-times.

Returning now to the energy used in UMSP. My main reason for this short paragraph is that UMSP cansometimes move massive objects such as heavy furniture which would normally require the efforts of two or three men, and there is an account in Lyall Watson's 'Beyond Supernature', of a four ton lift going up and down, through

several stories of a hotel that was being demolished, even though its electricity supply was turned off ! The phenomenon appeared only when a certain young member of the demolition team was present, and I thought at first in view of the weight involved, that the whole thing must have been the result of a collaboration of a hundred or more unconscious minds, belonging perhaps to people who lived nearby.

The heat energy given off by an average human being is about 70 watts so it would seem likely that the metaphysical/sexual energy consumed in a UMSP occurrence with one person involved could also be about 70 watts. This figure would be multiplied, in the case of the lift, by the number of minds (people) involved. We can work out the necessary power to be supplied by the people involved as follows:

Estimated height moved by the lift = 40 ft. Time taken = 10 seconds. Power used to raise the lift (ft.lbs/minute) = height moved (ft) x weight of lift (lbs)/time taken (minutes)

We can then write:

Power used = (40 x 10/60) x 8960 = (400/60) x 8960 = 59,733.3r ft.lbs/minute.

These figures suggest that far fewer people are necessary than at first thought, so, assuming that the number of people involved was 20 this would mean that each person on average would supply 2,986.6r ft.lbs/minute.

Since 33,000 ft.lbs/minute = 1 horsepower = 746 watts we can convert 2,986.6r ft.lbs/minute into watts by the following equation: power supplied by each person (watts) = 746 x 2986.6r/33,000 = 67.517 watts.

This is within our estimate of 10 to 70 watts/person and I think is a fairly realistic figure.

❖ 6. The bioplasmatic envelope (BPE)

The afterlife, as already stated, does in fact exist, and I have had many contacts with departed members of myown family and well-known historical figures, as well as a pet cat and a pet dog of mine which passed away a fewyears ago. The nature and medium of these contacts has been, in all cases, almost identical with that found inordinary telepathic contacts as experienced by me, except for one fraught occasion (several actually), when I sat on my bed aware in my mind of my then location but also of the point/place in the afterlife to which I was connected. I felt that the AL was above me and the current one was around me and that I was the single solid link, a tower of flesh and blood, which joined the two regions, although they would already be tenuously joined by the many mediums who must be, at any given time, talking with people in the AL via tenuous telepathic links. The other exception was that whereas, in normal telepathic contacts, there was only occasionally a visual component to the contact, in my contacts with the afterlife there was almost always a facial image, usually a remembered portrait. The mechanisms by which telepathy operates will be discussed further later. The actual discovery of the afterlife took place in 2005. I was pottering about in my room, pondering on this and that when I suddenly became aware, through the BPE, of my departed mother's presence; not as a person in the room but a consciousness of her identity and awareness of me, as I was aware of her. I was thunderstruck, amazed and delighted. My immediate thought was, "she is part of the BPE and must therefore be alive somewhere!" This, as it happens, is a perfect example of 'to feel is to know'.

During the following days I managed to contact her again, as well as my father and other relatives. I also made contact with many departed historical figures. One of the eminent departed with whom I made visual, but not verbal, contact, was David Hume the philosopher and he was sitting with a slightly amused expression on his face, as though he was thinking, "you were right then but I don't mind", knowing that I could see him and also that his being there
contradicted his refutation of metaphysics in his former life on earth. Earlier in the essay I defined the 'self' as tri-partite, that is:

(1) The essential self or spirit:
(2) The conscious mind which grows to accommodate acquired characteristics, (social relationships, education, everything learned since birth, personal tastes, habits, beliefs etc. and connects with the god within).
(3) The somatic self = the body.

I am quite certain that the spirit or essential self goes to the afterlife at the death of the organism and am equally sure that the rest of the metaphysical component, i.e. the conscious mind, as well as the GW, goes there too. The only part of the whole being which does not do so is the somatic self, the body plus its possessions. I shall list my other afterlife contacts later in the essay.

❖ 7. The BPE connects all life that exists or ever has existed.

The afterlife will, I think, be found to be much the same as the present one, in many respects, and the greatest difference, I think, is that one will not be encumbered by possession of a body, which means that physical discomfort and disease should be absent.. There can be little doubt, however, that it is not always a state of bliss, although I'm sure that such a state exists and may even be the norm.

The central truth to be drawn from the revelation of its existence, is that it is part of, included in, the bio-plasmatic envelope (BPE), which inclusion, in fact, was the cause of its discovery. Thus it is apparent now, extrapolating from the fact that it contains the departed souls mentioned, that the BPE contains and unites all life that is or ever was in a single all containing whole.

The objection may be raised that this is only a probability, but I would point out that the BPE is the result of BP amassed by the gods within, whose nature is to reach out through space and time and join, to a greater or lesser degree, so that ultimately all the gods within are joined and in contact, part of a global spatio-temporal web, and consequently so are all humans, and all other life-forms, even down to micro-organisms, which would not have individual divinities but would, as already written, be part of a general, tenuous intelligence or 'godliness' (or holiness, to paraphrase William Blake)

This fact makes time-travel into the past fairly straightforward though in a very different way from that described by physics.(This does not apply to time-travel in the afterlife). I shall be expanding on this and related topics later in the essay.

The actual logic that leads to the conclusion that the BPE joins all life that exists or ever did exist is as follows: The BPE connects all life on Earth, my mother is connected with the BPE, my mother died eight years ago but is still living in some form since I can contact her, therefore the place where she is existing exists. This place is, of course, the afterlife. It can be seen from this exposition that a tenuous mixture of BP and S, that is, psycho-plasma (PP), has been present on Earth ever since the first micro-organisms came into being, and it has been growing or shrinking in accord with the evolution of life ever since. In its later stages this general connectedness of all life through PP, has differentiated into the

CU and the CC. I think that BP and S will only rarely be encountered as separate substances above ground level but will manifest as PP. I think, however, that BP will be present and has been existent underground functioning as the CU to a depth of thousands of metres.

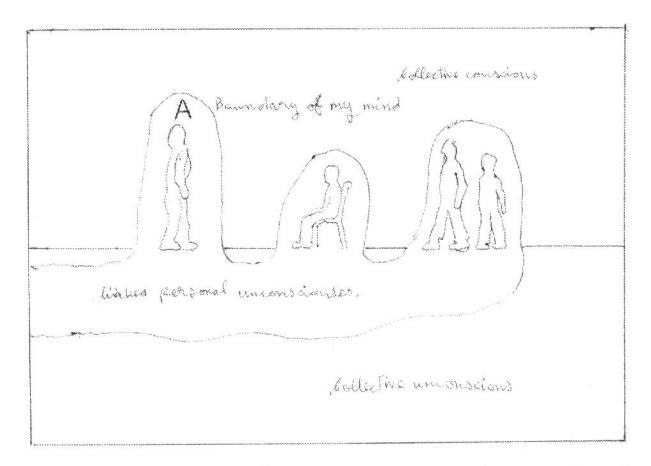

Fig.7. Conscious and unconscious minds joined and contained within a discontinuous boundary (effect of neuroleptic drug quietiapine ingested by A)

Unfortunately, although my knowledge of, and belief in the afterlife is absolute and well-founded, I have not yet found a way of demonstrating or proving its existence, although I do believe that the reality of Plato's forms as described in Chapter 5 item 6, comes very close to being supporting evidence that such a place exists, and it could, most certainly, also be seen as supporting evidence for the existence of the spirit. I have no doubt, however, that a watertight proof will eventually be found. It could be based on the principle 'To Feel is to Know', which I shall discuss later in the essay since anyone who has experienced the connections with the departed which have been granted to me, cannot fail to believe, such is the cogency of TFITK. (See addenda 115)

It could take the following form (and I recommend the reader to try this for him/herself). A person who does not believe in the afterlife, but is willing to consider evidence for its existence, is told that he can contact his departed mother if he is willing to try, (the mother/child bond is the strongest in the context of the family. In later life one may forget friends and sexual partners but one never forgets one's mother). He is told to find a quiet place and to go there every day for two weeks, to think of his mother for five minutes, her likes, dislikes, personal characteristics, smile, idiosyncrasies and memories connected with her, and generally be receptive to her possible presence. At first there will be nothing but if he remains placid and receptive it is possible that he may suddenly feel a contact and he will know, be quite certain from the feeling, that it is, indeed, his mother. This certainty is an example of TFITK and will be utterly unmistakable if it happens. If he gets no results he should leave it for a week, to rest and regain interest and motivation, and then he should try again.

NOTE. It has just now occurred to me that when a buddhist monk immolates himself as in chapter 2 (item 5b), his psychoplasma – the material part of his mind – might be destroyed as well as his physical body. I cannot imagine a self, a spirit, which exists with its attendant GW in the afterlife, minus BP which is always being amassed and deployed by the GW. I am sure that the departed would need his PP since that substance is his mind and personality. However I feel sure that a spirit robbed of its PP would automatically appropriate a fresh supply from the universal consciousness, which it shares with the PL. Putting this 'on hold' for a minute, I think I've finally seen how things must be in the AL. It is clear that, as stated earlier, the departed person or organism would consist only of PP and BP being in fact simply a disembodied mind, a mind without a body, god within plus spirit pluspsychoplasma. An imaginative description of the afterlife can be found under 'Addenda 127'.

Moving on now from this digression it can be observed that John Smith would occupy a place in the afterlife as he was at the time of his death. Telepathic access to John at any point can be gained by concentrating on a photograph of him at this point in his life, or if a photo is not available, anything which is associated with him. Rephrasing this in a rather dramatic but quite realistic manner: in theory I, the author of this material, can gain telepathic access to any presently living organism, and in the case of those that have passed on, any one of these, as he/she was at time of death; and many, but not all, of these telepathic contacts, could then be capable of being turned into teleportations, since all teleportations, I believe, will commence with the establishment of such a link. I say 'not all' for reasons that are revealed in the next few paragraphs. That is, I can, in theory, teleport any person on the globe into my presence at the computer here in my room, and moreover I can do the same with any person or organism in the afterlife. Picasso, Newton, Michelangelo, Leonardo, Aristotle, Tutankhamen, Neanderthal man, Proconsul and any of the early hominids from 4 million years ago in Africa, (I have actually been in telepathic contact with an early hominid. See chap. 8 item 6 entry 26) and this extends to animals, sabre-toothed tigers, mammoths and even tyrranosaurus Rex. (But only as they were at the times of their deaths, for reasons shown later). These later, more massive ones would probably take much longer to perform, (neglecting Bell's inequality theorem) though it will probably be decades, even centuries, before any actual resurrections will be possible. It is interesting to note that in theory long extinct animals can be recreated by using their DNA but this commodity is usually in poor condition, so the metaphysical method would, ultimately, prove much better as well as more humane. It is helpful to my theory to mention that the well-known co-author of The Anthropic Cosmological Principle, physicist Frank J.Tipler, in his book 'The Physics of Immortality,' affirms the possibility of resurrection but he arrives at his conclusion through the use of mathematical concepts in a universe of information processing.

I do understand that some of my claims may seem so preposterous, e.g. the hominid contact, that the reader may be reluctant to accept them, but please rest assured that I am telling the exact, unembroidered truth and am in full possession of my senses and faculties. It might be useful here, to refer the reader back to the quotation from the eminent biologist, now deceased, T.H. Huxley which appears just before the introduction, and the oneat the end of it, from Albert Einstein.

It has just now occurred to me that the fact that the recently mentioned John Smith can be contacted telepathically at any point in the AL at whatever age he died, relates very significantly to the fact (not proven but strongly supported) that time travel, by teleportation through PP, is possible. This requires more thought. I have also seen that an essential truth to be borne in mind when exploring these ideas, is that irrespective of its geographical or temporal location, there can exist only one of any particular person or spirit. I call this rule 'the individuality of the spirit' (IOTS). That is, there cannot simultaneously exist two Leonardo's one young and the other older. This idea was suggested by the supposition that one could try to perform a resurrection of Leonardo based on the picture of him as a young man, hoping that he would appear before our eyes as just that – a young man. At first I dismissed this idea; after some thought, however, I think this is perfectly possible having seen that the relevant ages or ***feelings of age or health*** in the parties involved are strongly determined by the nature of the relationship between the teleporter and the teleportee. That is, when I contact Leonardo as in the self –portrait of him with flowing beard he looks – and feels – like an old man and I sometimes feel that I am troubling him; but when I made contact with the younger Leonardo as in my picture of him as a young man he seemed very pleased and there was an excellent feeling of good humour and spirits. That is, the older Leonardo would be annoyed because I made him ***feel*** old, whereas my contact with him as a young man lead to abundant feelings of health, youth, warmth and good spirits. (Just short of doing a resurrection of him!). This kind of thing may occur in other places but it should be noted that Leonardo is a very large and special spirit. This supposition is supported by my own 'psychoplasmatic time travel' in which I almost regressed to age 14 (addenda item 42). I even heard/felt a slight 'plop' as Leonardo left his older self to regress whole-heartedly to his younger one.

In support of this judgement I would like to adduce an alleged case of travel into the future which I found on the internet at the following site: 'The Time Traveller, ACE OF SPADES HQ'. It is very cleverly staged and troubled me for months but, having total faith in my discovery of the individuality of the spirit, (IOTS), I felt very strongly that it had to be wrong. Accordingly, in my mind I went carefully over what I recalled of it while judging its veracity against IOTS. The alleged sequence of events involved a man crawling under the sink in his kitchen only to emerge minutes later into a strange place in the future, to meet an older gentleman who, it was claimed, was himself thirty years hence. The site showed a picture of the 'older self' taken on a mobile phone camera carried by the younger, and there was even a picture of identical tattoos, one for each body. This is in direct contradiction of my principle of the individuality of the spirit. The most satisfying aspect of this little mental excursion was that, as I pondered, an image appeared in my mind of someone putting his tongue out at me for refusing to be taken in by the hoax. The 'someone' was, of course, the person or persons who had perpetrated this little deception. I should explain here that in addition to the power of being able to contact a person telepathically, simply by thinking of them, I have developed other strange powers, one of which is the tongue-extending one, which can be interpreted as friendly and facetious but is usually a sign of annoyance and dissatisfaction, on the part of the person with whom I am in contact, or an accusation of being a 'spoilsport'. In any case its function on this occasion was to reveal, beyond a shadow of doubt, that this time-travel was fictitious. If it had not been a fiction I would have got, from the unknown creator of the site, a strong, friendly and triumphant feeling as I have on other similar occasions.

Changing tack slightly, I believed until recently that only the spirit went to the afterlife, and that the GW simply dissolved into the ubiquitous BP of the Universal Unconscious (which will be described shortly). I think it is quite certain now, however, that the GW, being a god, would certainly not fade away and moreover remains associated with its controlling spirit. Logically this must be so since my discovery of the AL hinged purely and simply on the fact that my mother was connected to the BPE via the GW. I have also modified my picture of how much of the organism goes to the afterlife. I have already defined the 'self' elsewhere in this essay i.e. the self consists of:

1. The essential self, the spirit as it was at birth, which has grown to accommodate the acquired characteristics, (the habits, beliefs, education, mannerisms, social status etc; i.e. the conscious mind and personality).
2. The god within and the personal unconscious.
3. The body or somatic self. This is the only part of the organism that gets old and doesn't go to the afterlife.

My second conception of the nature of the afterlife would appear to bestow almost unlimited freedom on any and every single organism to be found there. Any one of the higher animals would include a spirit and a GW in its constitution but free from the deadweight of the body, the spirit or personality would be able to direct itself with the aid of the god within to anywhere and, as written, anywhen, at any speed. It occurred to me earlier today that it would be useful and illuminating to try to imagine the afterlives of some of the non-human beings, who accompany us on our journey through life. A case in point is the fish and other marine life. One would assume that fish, dolphins perhaps, in the afterlife (AL), would stay close to home, and would not realise that they had died since they have, one would think, no conception of the AL. They would, however, quickly realise that something was different; they would be able, I think, to swim through solid obstacles, just as, I think, human beings in the AL would be able to pass through walls and ceilings (as do ghosts, and the person in an OBE who ascends through the ceiling to float above the city). This suggests that most if not all of the space in a particular city, or its streets, and houses, could be populated by the spirits of the departed.

In my first conception of the afterlife I assumed that the body would be reincarnated, but in my second one it is not. In my second one I also wondered why, when I contacted someone in the afterlife, I was aware of their identity, facial expression and general 'aura' or psychological type only, and could not see their environment. Today, while thinking on some other aspect of the AL, I realised that this was because the departed and the present life share the same environment – **_here and now._** The afterlife is all around us, in the kitchen where I made my discovery, in the living room and in the street outside – everywhere in fact! I also saw that the departed must be able to move swiftly from one part of the environment to another, my parents being a case in point. Sometimes they are very close and sometimes now (things were very different a year or two ago) they are far away in space or time. I am also fairly sure that they can see me and presumably also hear me, and my environment. (The spirit is responsible for perception). It is also very likely that any given member would tend to remain close to his friends and their joint locality in time and space.

❖ 8. The Universal Unconscious (UU).

It is also a curious fact that although conscious telepathic contact with the afterlife is at present given only to relatively few psychic mediums, the afterlife and the current life are strongly and permanently connected on an unconscious level, i.e. they share an unconscious which is the sum of the afterlife's collective unconscious and that of the current life. This is because, as stated, everything that lives or ever has lived is connected by the BPEand, stating this more accurately, the god within, of which each person and higher mammal has one, reaches out and is connected, via BP, to all the other gods in all the other humans and higher mammals, living or departed. Thus there exists a loose web of at least six billion (humans) plus x (higher animals) and y (lesser life) gods within, in the PL and an even greater number in the AL, all joined more or less closely in various ways, and capable of communicating among themselves. I call this entity, 'The Universal Unconscious' (UU) (diag. 3). I could have represented it as a straight line, upon which stands the current life and, an inch or two away, the afterlife, but portraying it as a circle is more satisfying aesthetically and more correct insofar as the AL and the PL share the same space and even the same collective conscious, not to mention the UU, and the circle I feel corresponds more exactlywith my recent perception of the permanent, ever present nature of past and present phenomena generally. I thinkthe future is unreal therefore it is not not singled out or indicated on this diagram.

This may sound like hubris, but to me the afterlife is as real as my fingers here on the keyboard of my PC; as real, in fact as Rutherford's knowledge of the existence of the atom. When I am in contact with it I can *feel* it, and to feel is to know (see chap.3 item 6, and addenda 107). My new conception has been described in other parts of the essay, and since its discovery I have realised that, since frequent conscious contact with it is relatively common, through mediums, they must be telepathic in nature. When the human goes to the afterlife, he/she goes in his/her totality – the only thing left behind is the body and material possessions.

Apropos of the foregoing, there must be a reason why the intuitive conviction of the afterlife occurs in literally every race and culture, apart from the obvious one of a desire to live forever. I think this reason must be the ubiquity of religion which is always associated with an an afterlife, plus the universal unconscious as just now described. . Wherever they are in space or time, a tribe, culture or civilisation will have access to information

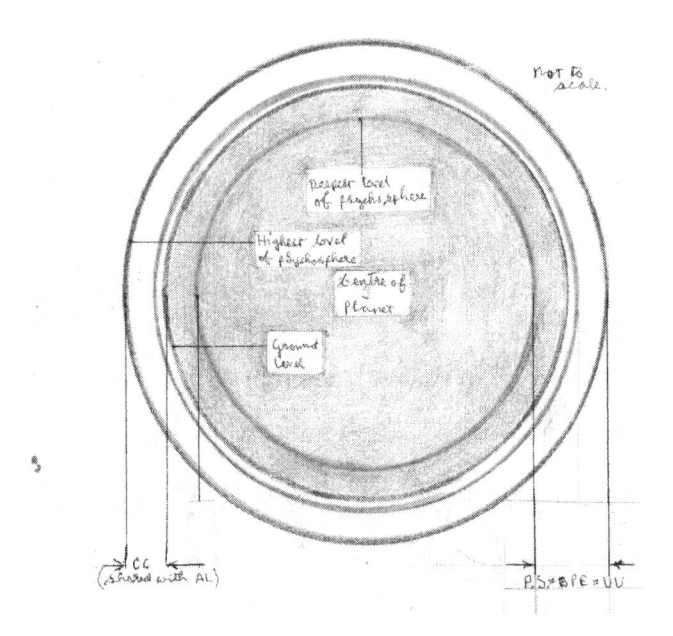

Diag. 3. The planet showing psychosphere, collective conscious and universal unconscious

being fed to them by unconscious telepathy, from the afterlife to members of the current life, and in doing so, granting them an intuitive perception to them that physical death is not the end.

It should be noted that the Universal Unconscious is identical with the BPE, and is ultimately another, illuminating way of looking at the same phenomenon.

❖ 9. Levitation.

I think the mechanism of levitation must be almost identical with that of UMSP, that is, closely related to the activities of the gods within, which, I believe, combine their forces to produce the phenomenon of lifting andmoving heavy weights and furniture and the lifting of the lift described a little earlier. The case of Daniel Dunglas Home in the mid-nineteenth century, who was observed to float out of an upstairs window and in again through another is very well known as are reports of 'yogis' in India performing the 'Indian rope trick'. I myself had a partial, (rapidly quenched !) levitation one day as I was lying on my back in the bath, but I managed to wrigglemyself down into the water again. I think the ability is related to the degree of 'goodness' or saintliness of the exponent (which immediately alerts one to the presence of the god within, which, being perfectly good, must be involved.). My own experience occurred during my period of mental instability many years ago, when I was living by an impossibly idealistic philosophy and being unrealistically kind and forgiving, never showing bad temper, irritation, greed, etc.; never offending friends or arguing with them, and I remember that at the time I was getting instructions, that is a little voice was saying, "feel the weight of your body". I was troubled all the time by a feeling that I was light in weight and liable to fly away, not comfortably settled on the ground. This state of pure goodness is involved, I think, in the phenomenon of divination which I want to discuss in my next essay.

❖ 10. Out of the body experiences (OBE) and consciousness.

The most interesting thing about these phenomena, to my mind, is that the spirit or aetheric body to use slightly anachronistic terminology, which stands by the side of or floats above the unconscious body, is fully conscious in exactly the same way that the waking person would be. (Ref. 9a). It is a fact, of course, that one is nominally conscious when dreaming, but one is not conscious of the environment and material objects surrounding one's sleeping body, as one is in an OBE. This, to be sure, must explain the mystery, or at least the origin, of normal consciousness. As the spirit or soul, it is primary and science has simply to accept this as a fact. One of the most elusive and pressing problems occupying science today is that of its origin and mechanism and neuroscientists all over the world are trying to find an 'awareness' gene, or other material biological component of the brain or brain/body, and of course they are wasting their time since it is a metaphysical thing.

It is quite clear that today's orthodox science does not admit of the existence of the spirit which, as just nowshown, is conscious when out of the body and independent of it, and in doing so, is relegating further enlightenment to the dustbin. Hidebound by this interdiction, research into the paranormal and associated disciplines is made forever empty and unrewarding, but free of its stultifying presence, the scientist can breathe in fresh air, and stretch his arms easily and freely into the vault of the heavens and the cool air of free enquiry.(Ref. 10a and 10b). It is true that in most cases the evidence for the occurrence of an OBE will only be a verbal report from the person who had it, and will therefore necessarily be construable as a figment of imagination but in opposition to that view I would like to adduce an experiment performed by the psychologist, Charles Tart, in the 1960s (Ref. 9a) which gives a clear, objective demonstration that OBEs really can be seen to be, at least, something extra to the body, leaving the body to float above it in mid-air.

I have before me another concrete example of the OBE (Ref.13). It would seem to be a small step to give this something the name 'spirit', something which I claim to be real and actual and, in view of the examples just now given, and the wealth of other evidence, I would consider as proven to exist. This latter example also indicates reliably that it is the spirit that affords perception to sensory information and not the body and brain. These are simply intermediaries between object and spirit.

Here then we have a clear, objective, empirical and quasi-replicable demonstration of the existence of the conscious spirit, and, I would think, a cogent argument for its status as the root of waking consciousness. The scientist generally does not admit to the existence of anything which smacks of metaphysics but I think he would find it difficult arguing his way out of this situation. Consciousness, at bottom, is a feeling of being awake and aware, and no amount of juggling with integrated circuits, AI systems, neural networks, computers and digital technology will create a system which feels, and no amount of dissection of brain tissue, or analysis of neuron feedback circuits will reveal how one can be created, because it is essentially a metaphysical phenomenon. In many if not most OBEs the subject reports an ability to direct him/herself to anywhere in the environment; e.g. into neighbouring rooms where she sees things, books, furniture, objects which the occupants of such rooms, strangers to the 'traveller', confirm to have been existing where she saw them. She may then float through the ceiling of the room she is in and the one above that to float freely above the rooves, and then to direct herself into different parts of the city where she's never been before. I think the mechanism of these events is very simple and obvious. The god within, naturally able to transcend space and time, is – possibly within limits- everywhere and everywhen. The possible limits would be, spatially, the boundary imposed by the surface of the psychosphere; and timewise the birth of the organism, as described in chapter 4, so that the spirit, the person's conscious self, need only direct its attention to its choice, a distant place, time or both, and the god within will take it there. It is worth observing, while we are on this subject, that as far as I know there have been no reports of OBEs in which the subject went into the future or the past. I must now correct this last sentence since I have recently read of an account where clear examples of this phenomenon were reported. (Ref. 17). I have also recently discovered evidence that time and space are practically non-existent in the AL so that the spirit can instruct the GW to take it to the remote past, possibly even, to distant geological ages.

These ideas, which may seem difficult if not laughable to the average adult, will be readily accepted bytomorrow's children whose heads and general mental apparatus will still be soft, pliable and malleable, accepting almost without question, unlike that of the fifty or sixty-years old who has to carefully compare new information with that already existing and verified in his mind.

❖ 11. Teleportation of an object, living organism or human being. I have experienced two complete teleportations and and two near teleportations of people, all of them spontaneous,and two spontaneous teleportations of inanimate matter i.e. a pot of Marmite and a guitar. The two near teleportations of a woman were of an ex-partner and on both occasions I had to abandon the proceedings for fear of harming her. I often find myself in telepathic contact with her although we haven't seen each other for twenty years or more. On one of these occasions, reported in my journal on 1/8/2006, I was sitting on the bed in my room, cogitating on this or that and I gradually became aware of the fact that I was coming into contact with her. The entry in my journal on this occasion was: 'Diagram illustrating the universal interconnectedness of the gods. I achieved this earlier in the year during a particularly crazy and amazing period (black light) when I was thinking of myself, having 'grown into the part', as 'master magician, metaphysician'. It was a truly astonishing time. I felt a bit like Dr. Faustus, the air in my room, in the evenings, heavy and dark in colour, loud banging and sawing noises coming from the ceiling and walls, and all sorts of crazy telepathic connections occurring. I was pretty f****d up and struggling for control to be honest. As already stated I almost managed to teleport (accidentally – I could have grabbed the chance), an old friend, a woman, right into my room, I even got a feeling of her leg coming into my room but I got scared, for her, not for myself. I thought I might harm her in some way. So I pulled back. Stopped it - feeling a bit like Dukas's 'Sorcerer's apprentice' who'd taken on a bit more that he'd bargained for but I had to keep going'.

On another occasion I was sitting in the dining room when I suddenly felt her presence and I just knew that If I was swift and sure enough I could whisk her into my presence in such a way that she would land lightly on the carpet like a fairy. Unfortunately I was too slow and clumsy and missed the opportunity. At that particular time I was much concerned with teleportation which is why I was getting so many spontaneous near-occurrences of it, I think.

It is a characteristic of teleportations, that is, it will be when they become commonplace, that their initial characteristics may be identical with those of telepathy. That is, for me with my powers, all I have to do is think of the name of the person I'm contacting and I'm in touch. A person not in possession of these powers would have to first think of his subject's name, then personal attributes of him, eye and hair colour, age, music preferences, hobbies etc. and would gradually, if successful, get a feeling of his contact's identity. At this point, the medium, or experimental worker, could encourage the feeling of identity and presence to go on growing so that the telepathic contact would gradually thicken up to become a teleportation. It is advantageous that the teleportee be well known to the agent, for if he/she is so there will already be a strong unconscious telepathic connection between them, which will facilitate the teleportation which is always initiated telepathically, or, in spontaneous teleportations starts with a telepathic contact I think. If the teleportee is a stranger the teleport. will start 'cold'and it could be hard work getting it going. This last paragraph is not based on any actual, successful teleportation done by me, but is a description of the procedure that I think would be necessary to accomplish one.

In a routine telepathic experience there will not be, I think, a large movement of PP between one mind and the other, but in a teleportation there will be a gradual, perhaps instantaneous removal and transportation of all the material of the body, including bones, teeth and enamel, the hardest substance in the body, for it is all suffused with PP. I have at other times been able to actually see the faintly sticky aura of PP surrounding and suffusing the teeth of a certain person in my past like pink candyfloss. It was quite dense, warm and glowing on her white teeth (she was a mile away from me but I could see it) but tapered off to fairly tenuous an eighth of an inch away. I have also, when shopping at my local Coop., got into difficulties and had to get out fairly quickly. When this happens I sometimes find that the warm cocoon of PP around my body (in other words my mind), gets stretched out so that I've left a large part of my head, in the form of PP, (not flesh and blood ! But still uncomfortable) in the shop ten or twenty yards down the road. (as shown in fig. 8). At times like this I have to stop, talk quietly to myself, and allow my mind to gather together again to surround me in the normal manner, before walking on. It is obvious that PP in the case of telepathy carries information from one mind to another, but precisely how this happens, as yet I do not know. PP must in some way, perhaps in its structure or subatomic configuration, if it is matter in the normal sense of the word, i.e. particulate, or some other mysterious way if it is continuous, (see addenda 109) be imbued with information, so that if a piece of it became detached from somebody's mind and lodged in a wall, or the ground, it could retain information which could become manifest in some other scenario, e.g. haunted houses or 'time-slips'. It would do this by becoming temporarily part of the mind of the person visiting the haunted house. I know that pieces can become detached because in the early months after having discovered that the mind was indeed material, it used to happen to me all the time because I hadn't learned how to handle my new knowledge, and when it did happen I found it quite disturbing as stated earlier. I am quite sure that ghosts do exist in the form of spirit and god within 'clothed' with psychoplasma.

At the turn of the 19th century Sir Oliver Lodge suggested that haunted houses could be the result of powerful tragic emotions which had somehow become 'recorded' in the walls of houses where murders or suicides had occurred. Half a century later a retired Cambridge don named T.C. Lethbridge came independently to the same conclusion.

Lethbridge had often experienced 'unpleasant sensations' in certain spots, as if something 'nasty' had happened there and left traces behind. (I owe this info. to Colin Wilson in his 'Beyond the Occult'). I think it ispossible that at the time of their occurrence the events' associated powerful emotions could have resulted in a piece of the mind (PP) of one or more of the people concerned, becoming detached in the excitement and being embedded in the bricks and mortar, as described a few lines ago, where it could remain for centuries.

Common sense says that in a controlled teleportation the density and volume of substance being conveyed (in the form of PP) should be inversely proportional to the time taken to effect the operation, but the ones that I have experienced don't seem to follow that reasoning, and it seems possible, by virtue of the GW, that instantaneous transmission is possible. It seems certain that voluntary, planned and calculated teleportation will become possible, having already experienced, at least, four spontaneous ones i.e. (a) the girl Jasmin who appeared in my bed; the reel of sellotape which passed through the doors of my kitchen cupboard; the the two young lads who appeared in my bedsit and the guitar which passed through intervening walls, doors and ceilings to appear in my room. My next move in this area, when I am ready, is to do some experiments with insects.

❖ 12. Dowsing and divination

These faculties are the ability some people have of locating the presence of objects, water, oil etc. hidden or buried underground. I am sure that they can be explained by the theory of material mind. That is, the collective unconscious and the lower levels of the personal unconscious are at all times present underground, in the form of BP, to a depth of hundreds, even thousands of metres, I think, perhaps even more, sufficient in any case to reveal the presence of deep oil and metal deposits. This supposition is reinforced by the fact that the neutrino, a subatomic particle, can pass, apparently,unhindered right through the planet. This being so, the GW, which senses, via BP I think, where the object, oil, coal or water is located, will inform the unconscious mind, of which it is the motivating force, and will also relay this information to the conscious mind by making the divining wand twitch or twist by a direct PK action on

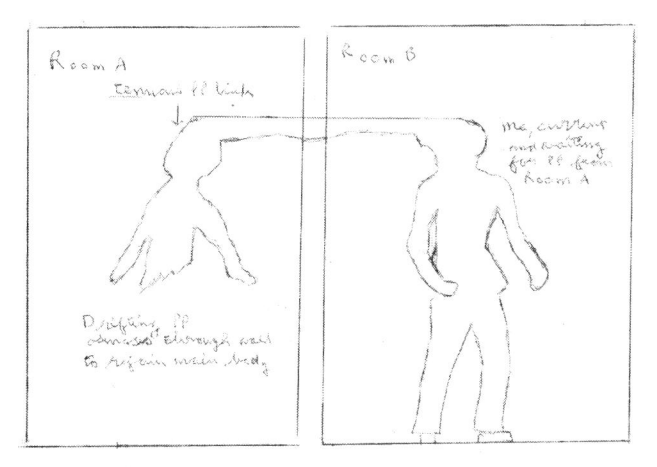

Fig. 8. Temporarily detached part of material mind rejoins main mass.

the wand, (not unconscious muscle spasms, as in the popular myth) which works via the dense BP that is certain to be present near the solar plexus, the home of the god within and which is where the wand is held (Ref.11e). This unconsciously generated PK will be similar to that operating in UMSP or poltergeist phenomena.

Despite this explanation, in whose truth I have perfect and total faith, it is a strange fact that dowsers cansometimes locate hidden water, oil or coal simply by hanging their pendulum over a map! I think I can explain this by referring to the fact, once again, that the GW, will know, via BP and the collective unconscious, where oil and water etc. is to be found; but it will also know, via BP and thoughts at head level, when the dowser is holding the pendulum over a place on the map that corresponds with one of these places, that is, deposits, and the pendulum will rotate exactly as in the former example.

Clarifying further, the method by which the GW obtains knowledge of where the pendulum is being held, isthrough the BP component of the PP that is the conscious mind. This is because BP is essentially the medium through which the GW communicates and obtains knowledge. Moreover, the means by which the GW obtains details of the places on the map will probably not, necessarily, be simply direct observation of the hand and the map, but, since there is an element of BP in the conscious mind, a direct knowledge of the conscious mind of the agent and what part of the map he is looking at.

CHAPTER 5

Extended youth and the elixir of life

1. The Anthropic Cosmological Principle; 2. Unqualified Atheism is no longer tenable; 3. The god within and true love invalidate atheism; 4. Freewill and the housefly; 5. Mental and emotional activity and psycho-plasma; 6. The dual nature of imagination; 7. Eternal youth - the 'elixir of life'; 8. Ghosts and their nature; 9. The conscious mind is finite; 10. The biological field; 11. Making a distant person turn round; 12. The non distance- -related nature of psi; 13. The Einstein/Podolsky/Rosen paradox.

❖ 1. The anthropic cosmological principle.
The Anthropic Cosmological Principle which is gaining credibility in various scientific circles, invites the supposition that the nature of the universe and human kind, are not simply the results of lucky coincidences but may be, at least partially, the result of some intelligent force or forces (one is instantly alerted to the presence of the gods within and, possibly, without) which may operate to provide a suitable matrix for the appearance of life. It is as though, not only are we adapted to the universe, the universe is prepared for our arrival,. (Ref. 9b). The mere fact that we are here, thinking about the universe proves that it can produce intelligent life. The question is *how?* This essay is, amongst other things, an attempt to answer this question. My purpose in mentioning it is to make it easier for established, orthodox science, which is almost always resistant to theories which threaten to change the scientific landscape, to accept my theory or, at least, to give it a sympathetic hearing before passing judgement.

The ideas behind another current scientific philosophy, intelligent design, also suggest, or rather, openly claim, that there is or are intelligent forces at work in the creation and maintenance of living systems. I myself would subscribe to this idea, and can state categorically that the GW plays an essential role in this, but totally reject the idea of a creator god or supreme being and organised religion, as anachronistic and an obstruction to progress. (Addenda 112)

These two growing ideas imply a new open-minded receptivity to the idea of mysterious divinities or intelligences being active in humankind's current picture of the workings of the natural world. The Theory of Material Mind is flooded with divinities and spirits good and not so good, all of which fit in, I believe, with a revolutionary new and emerging world-view which includes and unites physics and metaphysics to give an integrated and objective picture, as far as is possible, of the interactions of everything that exists.

Isaac Newton believed in God, but was an unwitting causal participant in a general move away from religious belief to belief in science and the scientific method. It is true that a section of the intelligentsia, Voltaire, for instance, continued to believe but theirs was a belief (Deism) based on reason and argument rather than intuition or feeling as too was Newton's. These latter were thrown out in favour of mathematical and mechanistic truths, i.e. a first cause was believed to be necessary to kick the 'clockwork universe' into motion.but right now the mood of the times is changing in step with the new astrological age of Aquarius, which, I am sure, as part of the larger scheme of things, is ushering in a rediscovery of this jettisoned knowledge. The truth cannot be buried for ever; vital forces, hidden pressures, will insist on its reappearance, like a dandelion plant buried beneath a pavement, growing and forcing its way through the asphalt back to the light of day. Although there is no direct relationship between the theory and astrology, (in which Newton himself believed), I have included in the addenda, as stated elsewhere, an 'a priori' proof which I discovered in 1975, that there is truth in the art and science of astrology, since it is a fact that happenings in the heavens march in step and interact with happenings on Earth as indicated in chapter 1, 'the interconnectedness of the universe'.

❖ 2. Unqualified atheism is no longer tenable

A system which does not possess freewill cannot be considered to be alive. If its behaviour can be wholly and exactly predicted time and time again to infinity, its status must be that of inanimate matter. My experiences indicate that a living system must include a region whose inner nature is beyond scientific investigation, and whose origin is necessarily a mystery. (Ref. 4b). A system possessing these attributes will be enabled thereby to possess at least a modicum of freewill. Creationism, of course, is a very naïve and insufficient explanation for the richness anddiversity of living things, but a purely mechanistic description of evolution and its functioning, (natural selection), though essential, is not sufficient. Newton was right in insisting that there must be an 'unmoved mover', that is, God, but the truth is that there are billions of unmoved movers in the shape of the gods within, their controlling spirits, and, extrapolating from this, possibly, other non-attached divinities and spirits. (See addenda 115)

Take a man who has been injured and is on the point of dying. His total being, according to the atheist, will be his body, the chemicals of which it is made, plus a mass of possessions, social relationships, activities, occupation, hobbies etc.. He dies……what is the difference now between his body as it is now he is dead and as it was a minute ago when he was alive? Attempts have been made to compare the weight of a body before death and immediately after. Apparently there was a distinct loss of weight, but the reasons for it were inconclusive.

So we must accept that physically and chemically there can be no significant difference except for the absence of pulse and respiration. His body is at the same temperature but something has gone. *He* has gone. His spirit, conscious mind and god within have departed for the afterlife the existence of which I am one hundred percent certain, but unfortunately cannot prove as yet, although I do believe that the fact that one can imagine perfect geometrical figures and conceive mathematical ideas etc. (Platonism) is itself a qualified proof. (See The dual nature of imagination: chapter 5 item 6 and addenda 115).

It should be borne in mind, though, that Hereward Carrington's attempt to weigh the spirit of a rat, produced a figure of 2.25 ounces which agrees with the TMM's conception of the material nature of mind.

Assuming that this is true, and that it is the whole mind that goes off to the AL, not just the spirit, one can make out a very good case for the supposition that the mind acquired gradually over the years, would have much more mass when adult than when it was a two-week-old baby. This may well be so, but it still seems probable that the pure, unadulterated spirit is of the nature of my art school imaginings, the ideal sphere, plane and point, like Plato's 'forms' which cannot be found in the real world of matter. These then would belong in the world of spirit, in other words it is not the spirit that is a substance but the mind, the 'acquired self'.

This reasoning clears up a problem that was raised by my second, and, I feel sure, correct, conception of the nature of the afterlife i.e. there seemed to be a contradiction between the idea of spirit (augmented by itsowner's mind) existing alone in the AL with its attendant GW, and my new conception of the AL complete with PP. I felt, incorrectly, that the spirit plus mind must be totally insubstantial in other words that the PP in the AL was superfluous and a disturbing inconsistency in my new conception of the AL. I see now that this is far from true; the apparently superfluous PP actually ***is the material mind*** which accompanies the spirit.

The foregoing three paragraphs can be summed up by the statement:

- A person in the afterlife consists of the spirit and associated conscious material mind (PP), plus the godwithin and associated unconscious material mind (BP).

This revelation also applies to the OBE with the difference that in such a case the spirit, or 'essential self', as I sometimes call it, and conscious mind, are still attached to the body, while the god within and the unconscious mind definitely remain immured in the gut.

At this point I would ask my reader to excuse the numerous self-repetitions and duplications which appear in thetext of TMM. I would gladly remove them but locating them would take too much time.

A human being is sometimes thought of by the atheist, who, it seems to me, must of necessity be amaterialist/mechanist, as an extremely complex, self-programming computer. But if that were so, the part that did the programming would still require a region of mystery – a secret self that could not be probed, and this mysterious region is, in very fact, the god within (GW), but this is not all. The GW animates the human body in tandem with the other metaphysical entity, namely, the spirit in the heart (S), which is the fundamental constituent of the conscious mind. It is the 'I' or essential self.

A digital computer which is what most computers are these days, is a system of logical 'gates', which for a given input of ones or zeroes (1 or 0) will give an output of 1 or 0, which quantities are manipulated according to the laws of symbolic logic (Boolean algebra, invented by George Boole in the 19th century) and binary arithmetic. An average computer will contain millions of these gates, each gate being built up of three basic gates which represent the logical functions of 'and', 'or', and 'not'. A computer will use a particular arrangement of gates, (and, not, nand, [not and], or, nor, [not or] etc.) to assess the states of its inputs, (what combinations of 1 and 0). Another arrangement will use a system of gates to count a series of operations. The maximum number of states of a digital computer, combinations and sequences, will be finite even though it may amount to billions. The quantities 1 and 0 can be used to represent 'yes' and 'no'. A digital computer, therefore, can at any given time, be in any one of 'n' states where $n<\infty$ which renders it completely predictable.

An animal on the other hand, e.g. a cat, is composed of atoms and molecules which are continually changing, changing position, combining, separating, forming chemical compounds etc.. It is an analog computer insofar as it uses continuously variable physical quantities such as fluid pressure, alkaline or acidic solutions, physical motion and so on. These continuous quantities form spectra going from min. to max. gradually instead of counting pulses as does a digital computer. The 'cat system' therefore, just like an actual cat, can exist in a practically infinite number of states which means that its state at any particular time is quite unpredictable so that it approximates more closely to a living organism than the former example.

Ultimately, however, discounting quantum effects' unpredictability at subatomic levels, it is still a network of logically clear, causal connections, involving no mystery or unknown origins. This system will be unpredictable on two levels:

1. Its status as analog computer.
2. Subatomic quantum effects.

Unpredictability alone, however, is not enough to define a living system. A random number generator isunpredictable, (discounting certain experiments in precognition), but it is not alive.(Ref. 10c).

One can only conclude therefore, that the god within, whose existence, to me at least, is incontrovertibly true, as I have felt within myself, read about, and observed at work in others, is the basic vitalising force in every individual organism on the planet, great or small, although I would imagine that in the case of plankton, mosses and micro-organisms it would be accompanied by spirit (S) in the form of a sort of nebulous, low-level intelligence which would animate areas of a few to hundreds of square metres existing as colonies. In higher animals such as humans, apes, dogs, cats etc. the GW is accompanied by a soul or spirit (animus and anima - male and female), which, in tandem with the head, directs its energies into the appropriate channels. Plants too, and all other life, have some kind of spirit which is probably distributed through their physical structure. Plants also have memory and can learn. (Ref. 6). One can only conclude from these observations that atheism is not sufficient to account for the functioning of a living system. Something more is required, namely, the spirit and the god within. (I would remind the reader here of Huxley's remarks about new truths and their inherent fate which is to be regarded at first, as heresies).

It would seem appropriate, at this point, to draw attention to the recent creation by Craig Venter of The Venter Institute, in America, of 'artificial life'. An artificial DNA nucleus was implanted in a living cell to produce 'science's creation of artificial life'. Specifically he has stitched together the 582,000 base pairs necessary to invent the genetic information for the generation of a whole new bacterium. It has been compared to Mary Kelly's literary creation Frankenstein although it did not entail the use of thunderstorms and bolts of lightning, but it promises to provide the means for the design and creation of an infinitude of novel and potentially useful life-forms, rather in the manner of robots except that they will be truly alive and, of course, when sufficiently developed, will be entitled to the rights which are appropriate to any higher organism. I am thinking here of such hypothetical creatures as the three-legged, (for stability), universal worker-stool which has four arms arranged round its central column (backbone) at 90 degrees to each other, atop of which is the head which carries four eyes, one for each arm, and the appropriate openings for intake and ejection of sustenance. This sort of creature will, obviously I would think, take centuries to develop and we shall probably have to be content for the time being with useful bacteria, slimes and moulds to do specialized work for us.

To be quite objective it should be pointed out that Venter has not actually created life, as he himself would be the first to agree, he has constructed a chemical chain (an artificial specimen of DNA) and inserted it into an already living cell. This is not to belittle his creation which has staggering consequences for humanity in the next few centuries, but merely to re-assert my own argument, in which I have total and absolute faith, since it has grown from both idea and experience, that wherever and whenever life is present so also are the metaphysical entities of the god within and the spirit. Moreover, as life-forms, Venter's creatures will be subject to the same laws and strictures as already existing higher life forms, that is, human life, and that of the higher animals. They will be part of the BPE and the universal unconscious and will die and go to the afterlife. Of this I am quite certain.

❖ 3. The god within and true love invalidate atheism

It must have been noticed by men and women cocooned in the gentle embrace of true love, that in extended periods of lying peacefully together in each others' arms, their heart beats gradually approach and settle into a state of perfect synchrony. This, it must be acknowledged by the atheists and materialists amongst us, is a powerful argument for the existence of a divinity or holiness inside every one of us. Not a creator divinity but a personal one. The god within of course, in partnership with the spirit. When god within and spirit are in perfect concord, and external, practical considerations affirm the beneficence of the alliance, the hearts of these true lovers will be completely open to each other; nothing will be hidden. There will be a dense psychoplasmatic connection at all levels and the hearts will beat as one.

❖ 4. Freewill and the housefly

A common house-fly can do things way beyond the capability of the most potent computer we have; this is not because of superior computing ability; today's computers work almost on a molecular level, quite comparable in active elements per unit volume with the brain of a fly. It is because it can organise itself and its activites, totally independently of an external supervising intelligence. It is independent and a 'law unto itself' unlike our largest computers which are necessarily programmed by human intelligence.

This independence is also not due to superior computing ability – this is actually only a tiny fraction of that of the average laptop. The housefly's independence of a supervising intelligence is because there is a metaphysical dimension to the fly, a soul or spirit and, (a slightly comical idea, but true I think) – a tiny god within, or, perhaps, a share in a larger GW which could encompass several organisms, since the fly possesses a thorax and abdomen, just as in a human. This gives it freewill which is an essential prerequisite for declaring a system alive or constituting a life form and therefore not completely predictable. Even the tiniest microorganism must possess a modicum of this commodity even if only an infinitesimally tiny and undetectable measure.

❖ 5. Mental and emotional activity and psychoplasma

The basic stuff of the mind, bioplasma (BP),originates at cell level, I believe, through-out the body, and becomes multiform feelings and sensations in the gut, is concentrated and amassed by the god within in and around the solar plexus. This material combines with spirit (S) in the heart, where it is organized by the spirit working in tandem with the god within, into words, images, ideas and complexes of various feelings. This organizing and playing with mental phenomena, although based in the heart, occurs to a greater and more controlled and refined extent in the head and outside the head to a distance of two or three metres in normal thinking, but can happen at even greater distances when necessary, (see following item). This phenomenon is illustrated but on a much larger scale, by the popular expression 'building castles in the air'.(fig. 9).

The basic psychoplasmatic activity which is always present in the heart, is what is known as intelligence. I see imagination as a function of intelligence and I think that understanding is derived from the latter, although I'm not fully acquainted with the relevant and current philosophical and psychological thinking on the subject. All of these activities originate in the heart, as does much if not all conscious mental activity. This material in its totality, including information from the senses and the body itself, energetic or exhausted, warm or cold etc. is sorted, ordered, analysed and classified in the head, the cognitive centre, and, where appropriate, despatched to places in the heart, upper body and gut where it is stored as memories, regulates biological processes or is expressed in physical actions .

It has in fact, been found recently, that memory relies upon and operates by means of neuron feed back circuits, of which many are found in the heart. This research is quoted more fully elsewhere in the essay with the relevant reference. This supports my theory insofar as the theory states that memories are located in the heart, abdomen etc. not just as memory traces in the brain. The head, in a properly individuated person, (individuation - a Jungian concept), sits atop the 'pile'of dense PP in the heart which is held together when danger from a social connection threatens, by a hunching of the shoulders and protection of the heart region by the crossing of the arms across the chest (self-protection). This also holds together the material, PP, organized by the spirit in the heart, which although centred, concentrated and originating there, (the heart), would extend into the surrounding space becoming more or less dense depending on distance and its degree of connectedness to other organisms.

The heart can operate in an 'open' or 'closed' state, (e.g. 'open-hearted' or 'close to the chest'- 'cagey'). In most situations the head will be in control (that is – oneself, wary and alert, and present mainly in one's head) and will regulate the open/closed state of the heart. In an act or general state of receptive openness and awareness the heart will be 'open' to the passage of information inwards or outwards, in the form of PP. This state would most likely exist at home with one's family from which one would have few secrets, and whose presence constituted no threat, or if one were unburdening oneself to a 'bosom friend'. The heart would, in the norm, be fairly well closed in the company of strangers, but would be capable of any degree between open or closed depending oncircumstances.

❖ 6. The dual nature of imagination (DI)

Nothing is, or ever was, entirely imaginary or non-substantial. (neglecting, that is, the god within and the spirit ofwhose essential constituents we are ignorant and probably always will be). I recently discovered that what was formerly considered to be unreal, i.e. imagination, consists of a continuum of PP, going from heavy and viscous in the heart to clear, light and rarefied in more distant places. This means that if one tries to imagine something, let's say a teacup, there will be, at some nearby point in the space around one's head and/or heart, an image or model of a teacup which, far from being non-existent physically, will be a definite material object, consisting of course, of psychoplasma, possessing mass, although extremely tenuous, and a degree of solidity.

This applies to all 'imaginary' phenomena of a conventionally real nature, that is, any entity such as a table, dog, tree, a boxing-match, a thunderstorm, a fire, melody or any material object, and such things are usually, though not always, conceived as being outside the body in the surrounding PP although the subject may not normally be aware of this. They may also exist at any size between much smaller (inside the head) to much larger than the actual size of the object(s) imagined.

Imagination in some form is employed in practically every human activity. Let's consider the case of putting up some bookshelves for a friend. I am in her house. I think, " where is my screwdriver?" And an image of it floats into my mind. It will be constituted of psychoplasma as will the succession of images accompanying the activities of lifting the shelves, holding them in position and screwing the brackets to the wall. This PP will suffuse my body, and cling around me as I work. It will be quite invisible to the average person. As explained, this sort of imagining will accompany most human activities including getting on the bus to depart for work in the morning, i.e. I will see the bus in my mind, probably before it arrives, the people at the stop, and then, as I sit on the bus, entering a much denser region of PP, possibly visualise the office or studio where I'm going. All this too, will be composed of PP at various densities, and on smaller or greater scales. These are all examples of material imagination which is related to 'the reality of matter' as dealt with in chapter 2 item 5. At the same time I will join the mass of PP already on the bus, absorbing some of it, rejecting some and also making my own contribution.

Now I will bring forth certain playful imaginings in which I indulged when at art school. While drawing and modelling clay in first year I often found myself creating mental models of geometrical figures. One of them was of a sphere of clay pressing gently against the studio window pane, and becoming slightly flattened at the point of contact. I then imagined it as being an absolutely perfect, transparent sphere, touching an absolutely perfect flat surface and to my delight realised that neither the point, plane or sphere could exist in material reality, but were simply collections of dimensionless points. The point at which the sphere touched the plane would have no spatial dimensions but would simply be a place, and the surface of the sphere would be an infinitude of places, as would the surface of the plane. So the sphere and plane could not exist in material reality, since, if they did they would be made of particles, atoms etc., and would therefore be vague and/or knobbly at the edges, that is, imperfect. This applies to all geometrical figures triangles, squares, etc. and to numbers, which must be perfectly exact. In the material world nothing is perfect. But the fact was that they clearly did exist somewhere and that somewhere was my mind - its spiritual part. And the reason for this was that my conscious mind is based in spirit, not matter. If it were based solely in matter as current neuroscience assumes, ***it would not be possible to conceive a perfect sphere or a perfectly straight line for matter is made of knobbly particles and fields.*** The root or origin of the conscious mind is the spirit in the heart, and the spirit is capable of leaving the body in the OBE (while leaving the god within in the gut). The clay sphere, before conception of the perfect, transparent one, would remain simply as a tenuous, spherical volume of PP and therefore material.

It can therefore be seen that what formerly was considered imaginary, unreal, of the same nature as the ideal forms, is in fact material, that is, made of psychoplasma.

Now if I can summon up in my mind, a perfect sphere, a thing which cannot exist in the ordinary, real world, it must be possible for John or Bill to do the same. The big question now is, is the ideal sphere in my mind identical with that in John's and Bill's? – put plainly– is it the self-same sphere? It is not and the reason is that the spheres of John and Bill could be resting lightly against different windows from that of mine, and if the windows were different then so would be the spheres. I was going to say, wrongly, that perhaps the two – or three – spheres, since they relate to and are functions of the spirit, may be identical and existent only in the afterlife since that is the realm of discarnate spirits, but then I remembered that my second, truer, conception of this region, the AL, placed it exactly congruent with the space in which the current life exists, which implies that the space of the AL is the same space as that of the PL; the spirits of the departed are all around us, or so I believe (see addenda 110). Referring now to diagram 1 it becomes clear that the individual spheres of Tom, Dick and Harry each occur in the relevant private part of the universe, while the public universe remains accessible to all three, and people in general, including people in the afterlife.

So how does this conception fare in relativistic space? The GW and the spirit of the departed being would be mass-less, I think, which suggests that BP alone, which I believe to have mass, would be subject to gravitational effects. The departed organism as a whole would be subject to exactly the same forces as are the living, but the spirit plus its associated BP, as in the OBE, (ref. 9a) could travel, at varying speeds, perhaps instantly, to anywhere in space-time carried by the GW, and thereby transcending the limits imposed by relativity. (See chap.5 item 13)

The paradox is that when one imagines these ideal entities i.e. the perfect point, plane and sphere, one is imagining nothing – there is nothing there – although there clearly is – an immaterial form (or place, in the case of the point, to quote Russell). We must remain content with Plato's descriptions of them as ideal forms, each existing in a nonmaterial world and thus giving at least a modicum of validity to my assertion of the reality of the spirit.

Psychoplasma is a mixture or compound of spirit and bioplasma, the latter being secreted, I believe, in the cells of the body in humans and animals and in vegetation where density of psychoplasma D = 1 and its 'basal density' below which, I think, it will not fall anywhere on the planet, could be in the region of 0.001 (see graph 3).

Such things as smells, tastes, colours, sounds, tactile phenomena, are perceptions and are mainly internal things although they may partake of externality and when imagined, are effectively perceptions manufactured by the spirit which, I think, is responsible for all perception. The ideal forms are, I think, essentially mental if not cerebral; they are conceived at or about the forehead while the god within and BP are essentially visceral.

When one looks across a sunny landscape to the distant mountains, one is looking through several miles of tenuous, rarefied psycho-plasma. This is one's conscious mind and the CC into which it fades. The mind is rooted in the heart and thorax where it cooperates in a dominant or leading role, with the personal unconscious motivated by the god within. It extends into the surrounding space becoming progressively more tenuous until it reaches a lower limit, unless it encounters another mind with which it may merge. In my diagrams I have graded PP density, as written, on a scale of 0.001 to 1 (Rhines?) one being the thick, heavy, smooth and densest state which is in the heart, and one divided by one thousand being its lowest tenuous (basal) state in distant unpopulated regions; this will probably be fairly constant due to the multiplicity of life forms generating PP at any one place, and the geological ages during which it would have had time to become homogeneous and settle into its current background density. As written, PP is densest in the heart and is formed in the body by BP from the unconscious mixing with spirit (S). It is exuded finally through the skin to form the mind of the individual which itself will eventually merge with other minds and/or the collective conscious. This can be succinctly stated in the following manner Nothing at all in the conscious mind can be immaterial or insubstantial except the spirit, which I believe to be an immaterial essence of the self or possibly a region of pure awareness. The dual nature of imagination has ramifications such as PK, teleportation of a living being, or inanimate object, both of which which I have witnessed, retro-travel in time, 'eternal youth' and more. Each of these phenomena relies on the material aspect of imagination for its operation. Although I have not yet performed an intended or pre-planned teleportation - though I have encountered spontaneous ones – or experienced retro-time-travel in the PL, I have come so close to doing so that I feel perfectly sure that such things are possible. Each one of these phenomena relies on the material aspect of imagination plus the motivation of the spirit for its operation. My next move in this area is to try teleportation of insects, which should be easier than bits of inanimate matter, although that has been seen to be possible, since insects can be presumed to have a metaphysical component in their makeup. This may entail cultivating 'pet-style' relationships to warm things up a little. (I heard somewhere that Ovid had a pet fly, which reminds me that I myself almost formed a material bond with an earthworm [chap.7 item 7] and even formed one –material – not just psychological - with my houseplant!).

Returning now to the subject of DI, let us imagine a teacup. The image will probably appear within two or three feet distant from the head, although the reader will probably never have given thought or been consciously aware that items in the imagination do, usually, occupy a position in space relative to the head as opposed to inside the head itself. The image will be very tenuous but probably not so tenuous as the background PP of one's mind and the even lower tenuity of the CC, but it will possess substance. If its only quality is its spatial form it will be, I have found, a translucent light amber tending to clear and transparent in colour. Now let us imagine it as a piece of bone china. It may acquire weight, colour –brilliant white –and one may hear, 'imagine', the tinkling sound produced by tapping it with a pencil or similar object. One might then imagine coloured pictures of country scenes on it. Is the colour real, made of paint, or some other substance? Or a perception manufactured by the spirit? I feel pretty sure that it will be the latter, and also that the spirit is responsible for all perception (see chapter 2. item 4). My reason for thinking this is that in a teleportation, I believe, the material to be teleported is dissolved completely into the form of PP to facilitate its moving through space to its destination, unless it is instantaneously transported, which may be possible (see ref.16 and chap.8 item 6 and 56). This seems logically coherent and feels right, but raises the observation that there are four dimensions or aspects to imagination, i.e.

1. Spatial imagination: the creation by the spirit of 3-D forms in psychoplasma.
2. General or compound imagination of qualities extra to spatial imagination: colour, texture, temperature, mass, transparency, hardness or softness, etc..
3. Abstract entities: beliefs, conversations, ocean cruises, melodies, holidays etc..
4. Concepts - or universals - as in philosophy: Wisdom, justice, punctuality.

As an illustration of this let's try imagining a bar of chocolate. It will appear in our mental field a foot or two before our eyes, outside our bodies, and if it is imagined with its spatial qualities only, it will appear as a clear, translucent block. If it is imagined in its sensory totality, however, it should possess form, dimensions, texture, colour, smell, taste, weight (mass) etc. One could ask at this point, since the mind and all its contents are material, how real would the bar of chocolate be? And how dense? Would it be composed at all of milk, sugar, cocoa even if only minimally? When one imagines the taste of sugar one can sometimes almost literally taste it. Is one's tongue involved? Having just now, as I type, imagined the taste of sugar, the first taste occurred in my back! The following experiments however, registered on the tongue and definitely tasted slightly of sugar. This was, of course, a memory of preceding perceptions.

When I tried a few moments ago to imagine the tinkle of china I was definitely aware of a quasi-sound or tinkle that I heard or felt in my head, neck and heart. I think the conclusion to be drawn here, is that the spatial image will possess the appropriate spatial properties, it will exist in real space and have dimensions. The other qualities such as odour, taste, colour, texture, mass will be examples of perception remembered by the spirit. In a teleportation, gradually (possibly instantaneously, see ref. 16) transferred, perhaps in a very tenuous state to B, where it materializes. This suggests that the mind can convert and contain, in the form of PP, the material qualities of the teleportee, that is, flesh and bone, dental enamel, textiles, jewellery, etc.. This is confirmed by my experience with the Marmite jar and Sellotape, in chap.8 (item 6 no. 24. 25.), 'Author's Experience of Paranormal phenomena', and chap.8 item 56 when a guitar passed through several walls and/or ceilings to appear leaning against the bookcase in my room. This is clear confirmation that inanimate matter can disappear at (a) and simultaneously reappear at (b). It may also pass through other inanimate matter i.e. textiles, jewellery, as well as living bodies. Pondering on this, I think that such conversions of matter into psychoplasma must always be complete so that the mind will always be composed of PP alone, and not differing substances.

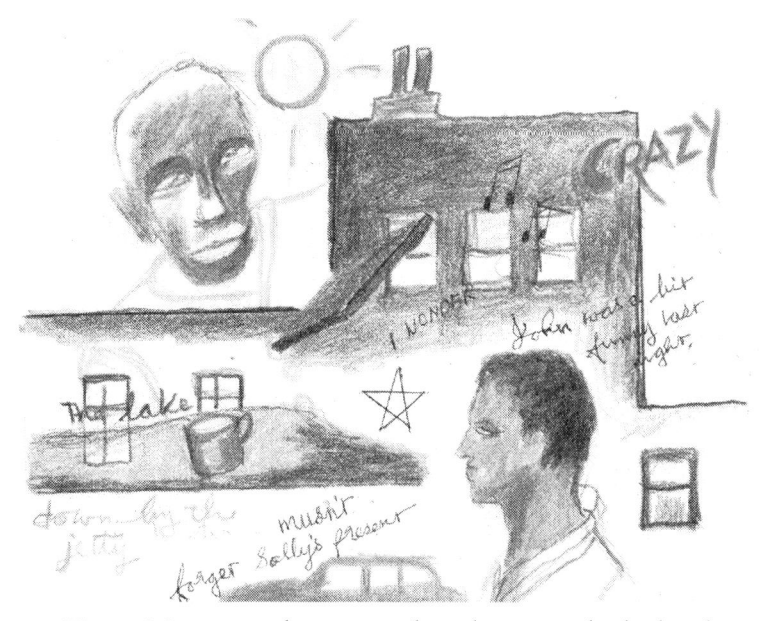

Fig. 9. Most mental activity takes place outside the head

(It may be germane to the issue to mention that sometimes, when I am stressed or exhausted, I notice a very bad odour from my bathroom sink even though it is cleaned every day. It is very strong and never happens when I am in good spirits. It is clearly an olfactory illusion or perception manufactured by the spirit – my spirit - in an exhausted state).

I should remind the reader at this point, that although I have not intentionally performed a teleportation, I have come so close to doing so that I am practically certain that such a thing is possible. I can also cite the occasion, already mentioned, when I woke in the night to find a strange girl in bed with me, who flew off through the wall near my bed. This, of course would be classifiable as a spontaneous teleportation. (Chap. 8 item 6 no. 9).

What must happen then, when one performs an act of imagination, I think, is that the spirit, in tandem with the head, being the essential organ of perception, among other things, operating at head level, creates a perception or perceptions of the imagined matter independently of any cause or stimulus. This perception will be 'clothed' in the circumambient PP as are all products of imagination except the Platonic forms. That is, unlike most ordinary mental phenomena which originate in the heart and are manifested at head level, the ideal forms originate and appear only at head level in the surrounding space, I think.

It is clear that in an ordinary, day to day act of imagination, let's say planning the day ahead, all the contents and actions of the mind will originate in the heart (spirit) although they will be felt and operated on at head level. They will be composed of PP at various densities, plus a constellation of words, objects, people, places, times and actions, which will become manifest at, or centred in the head, and occupying the space two or three metres above and around head level where the density of PP will be much less, but more capable of manipulation into words, images, feelings, processes and general mental phenomena (fig. 9).

❖ 7. Extewnded youth (the alchemists' elixir of life).
One of the more interesting consequences of the DI is the possibility of extending one's youth; not just extending one's life-span although this would be a parallel consequence, but preserving the feelings of youth, and consequently the 'bloom of youth'. This is possible by virtue of one of the revelations that came to me in my lonely period, i.e. 'the inner truth is greater', meaning that the spirit, and consequently the mind, is stronger than the body; (see chap.2 item 5b), therefore with the growth of increasing years, as the body ages the spirit should not press it too hard, thereby giving it time to recuperate, regenerate and gather its resources ready for another period of youthful activity. Thus the body need not become old and tired but can simply slow down a bit, and take a rest, while the god within supplies its unlimited funds of new, youthful feeling. ***It is the body that ages, not the spirit.***.

This would probably also involve concomitant and consequential reparation or cessation of cell atrophy by the subatomic operations of the god within. The influence of the GW on subatomic particles is strongly suggested by the fact that the functioning of electronic equipment is influenced by it as shown in chapter 8, 'personal experiences of psi by the author' (chap. 8 item 6. no.16). The GW and the spirit are agents and the former is constant and absolute; the body is malleable, and susceptible to their actions, and therefore variable; the spirit too, though being essentially an independent agent, grows in parallel with the body to become larger and more complex. (See 'The artlessness of youth. chap. 2 item 3). Oddly enough I had a most excellent experience of being twenty years old again this very morning. I am well past fifty and much of my habitual thinking is centred around the desire to remain young, so any experience that comes my way which relates positively to this is happily accepted. This morning's experience was not simply one of feeling alive, motivated and full of energy, which I was, but one of being the person that I was at age twenty, and I felt it flooding through my heart, lifting me and inspiring me to keep up my mind-blowing researches. The fact is that the mere knowledge (or necessary belief), that it is possible to extend one's youth, will make it possible. If one understands this or strongly believes it, all one needs is strong, sustained drive to get through the difficult bits that occur in most peoples' lives, to emerge into the freedom and sunlight of carefree youth again. And it is not just a feeling of being younger – the whole system, physical and mental is affected. Hair should grow darker – new hair – teeth may grow whiter (not sure here but probable) – and gums grow young and strong again. This last one I experienced while using my 'cross-trainer' exercise machine. As I pedalled I suddenly realised that this feeling of youthfulness was with me again and I actually felt my gums swelling to cover the roots of my teeth. It was as if a large, powerful but benevolent being was gently pushing me forward into a new, renewed state of being. (Writing now a few weeks later it has become clear that my teeth *are* growing whiter!)

This is where TMM accepts but bypasses or cancels the normal, recognised fact that too much stress tends to make one feel and look tired and old. Acceptance of the truth that the GW is responsible for amassing PP and generating new feelings of youth, energy and growth automatically bestows these three things on their recipient. It goes without saying, of course, that he/she must, of course, stay in touch with his feelings and not over-exert but accept a gentle, gradual return of youthful wellbeing.

Here, oddly enough, unlike as in orthodox science, one must believe not only when the evidence supports the hypothesis, but even when it flatly contradicts it ! One must have faith, not in any creed or religion, but in the power of the the mind and the god within. When one sees the lines and wrinkles in the mirror, one should not think, "I'm getting old", but "I'm overdoing it – better slow down for a bit, I need a rest ", and soon one should be feeling better, becoming aware of oneself and one's surroundings again, ***waking** up!* Out of the long, dawdling dream into old age.

There is a saying that 'when the going gets tough the tough get going', and this most certainly applies when one is trying to preserve one's youth. This may sound like a contradiction of what I have just now said about 'slowing down for a bit' when the wrinkles appear, but this is not so. What I mean is that when everything is against one, and life starts getting hard, one may start thinking that one's feelings of weariness are evidence of increasing age, and one may feel like giving up, and creeping off somewhere to die, but this is where knowledge, and if this is not available, faith, in the power of the mind and the GW, is absolutely essential to keep one going. If one can do this, even if only maintaining the status quo and taking one day at a time, one will eventually start to see and feel new, youthful sensations, glad feelings of lightness and joy and before long one is wholeheartedly believing the evidence which is now totally positive. I personally, having recently emerged from a very trying period, fighting off the negative thoughts engendered by difficult but unavoidable social relationships, am now reaping the rewards of a dogged drive to keep going whatever the odds, knowing as I do, that no matter how tough life can get, if one keeps going one can emerge victorious, shaking off the years along with the bad feelings. This is an oscillation that I know well having had a few of them myself and each time that I 'surface' into a younger stretch of my life is better since the mechanism grows clearer and clearer and the knowledge surer and more certain. I think that it is safe to say that this process can be extended to well over one hundred, and even one hundred and fifty and further. We are now living in the New Age of Aquarius and almost anything will be possible, I think.

The whole process of retaining one's youth is largely a matter of belief as was the achievement of the four-minute mile by Roger Bannister in 1954. He achieved a time of 3 minutes 59.4 seconds and shortly after the 4 minute barrier was broken again by Chris Chataway and others. This despite the fact that for the previous hundred years or more it had become the general belief that it was impossible. The four-minute barrier was, after all, a psychological one, as is the current allotted life-span. In 1953,the year of the coronation of queen Elizabeth 2nd, Everest was climbed for the first time and the dynamics of that achievement were similar. I would like to write more on this subject but may postpone it for my next essay. Meanwhile, **GET THAT FEELING** of youth.!!!

❖ A distillation of the foregoing.
If you will maintain, regain or extend your youth, you must believe – or better still - *understand*, that such a thing is possible (remember bannister's 4 minute mile) in your conscious mind. Ignore the negative evidence –The GW doesn't bother with evidence. It's power is practically unlimited but it must obey the conscious mind. And it is the GW that supplies the unlimited funds of good, new and youthful feelings so one mustn't suppress it by becoming disheartened and refusing to believe. The GW lives mainly in the gut but is strongly present all around the pelvic region and down the legs. As an artist would say; if you will create, (or become young again) turn off your head, thus liberating the enormous powers of the GW and the unconscious, or, if you *must* think and analyse, at least use your head properly, to eliminate prejudice and to construct and reveal positive knowledge and belief.

I believe that all depression, contradiction and lack of belief occurs in the conscious mind. The unconscious one motivated by the GW is a permanent fund of optimism and energy, despite, as alleged by Freud, Jung and others sometimes containing disturbing and repressed material from the conscious, which, I am sure from personal experience, eventually withers away if a strong, positive and optimistic drive is maintained.

AN OBSERVATION: It is in the nature of the conscious mind, based in the spirit, to fight others, contest, survive ……. or succumb. The gods within, however, are friendly, gregarious, helpful and tend to congregate and co-operate. The spirit generates barbs of wit, the god within generates humour.

NOTE: It is a well-known fact that the mind influences the physical state of the body (psycho-somatic illness and its opposite) I call this 'intra-corporeal PK' since the operative energy comes from the GW – it is essentially the GW acting on the body. All 'negative growth' or negating evidence is the work of the spirit or conscious mind which must think itself free of these constraints.

Since we are still on the subject of quasi-immortality I feel obliged to relate my most recent experience of 'waking up out of the long, dawdling dream into old age'. It happened about half an hour ago and was caused by the interval music and its accompanying picture on BBC4. This strange, enthralling music and the slow turbulent heaving of the waves of the sea as they crashed on the rocks took hold of my heart, which was young again, and I felt myself lifted and feeling, at the same time, a youthfull strength and power spreading deep inside. I also had, again, a powerful sense of 'waking up' but this time it was not just my head waking up, but my heart – my whole being, and I felt connected once more with the 'spirit of the age', the whole vital, gamut of current goings-on. I was back in the thick of it, and it felt miraculous. The whole thing lasted only twenty seconds or so but happened again, though slightly reduced, every time I heard the music. There is no doubt that I am gradually regaining my youth and extending my life, after a period which would probably have killed someone lacking my knowledge and conviction.

Finally, it might be asked by the sceptic, 'why do animals grow old and die?', and one would answer, "because they respond only to their feelings, and have no way of escaping their reality and power, whereas humankind has intellect and consequent knowledge i.e. science, by the use of which it can sort out positive feeling from negative, and fact from falsehood".

For me, this whole sphere of retaining or regaining one's youth is closely connected with thoughts which came to me, such as, 'all gods are good but some serve evil ends', in the extremely lonely and alienated period of which I have written earlier. (See chap.2 item 1). These thoughts, which I trusted since they came from the spirit itself, (or so I believed at the time) in other words *me*, were 'the inner truth is greater', and it took a long time to ascertain exactly what they meant, and where they were to be applied. Seen from my present position of far greater objectivity vis-à-vis my discoveries, I have no idea where these thoughts came from, but see that it was right to trust them; it is also clear that they refer to the independence of the spirit as compared with the lesser independence of the body. (See chap. 2. Item 5) Having seen this, it is easily seen that the independent, believing spirit should be capable of taking charge of the entirety of the body, and optimising its activities so that they relate more flexibly to the variable demands of the environment, thus enabling the body's survival.

❖ 8. Ghosts and their nature.(most tenuous at PP = Basal density – quite invisible)

It is interesting to consider the nature of ghosts, of whose existence there is no doubt. Ghosts, I think, must consist of PP at or near basal density (normally invisible), which has somehow acquired a modicum of additional PP thus acquiring also a degree of materialisation. When one is seen missing part of its body, legs and lower trunk for example which sometimes happens, the missing part must remain I think, as PP at basal density.

It is commonly reported after sightings of ghosts that there was a fall in temperature when the ghost appeared. I think this must be a result of the fact that the ghost would be composed almost entirely of extremely tenuous PP and would almost certainly suck warm, body temperature PP from its surroundings and observers, thus reducing the density of warm PP in the room resulting in a fall in temperature.

This phenomenon, with one exception, is readily explained by TMM. Each one of us is immersed in the pool or blob of psychoplasma which envelopes the group to which we belong, and which in its turn, is part of an even larger blob enveloping and connecting it with the collective conscious and through which we stay warm and connected. The ghost, when it appears, will probably have come from some anomalous situation in the afterlife, in which it was not closely connected with its own group PP-wise and was therefore manifesting as a cold, possibly lost and straying, spirit. This is largely conjecture but it seems reasonable to assume that it would be cold due to a very low or tenuous PP content so much so that when it appeared to spectators a proportion of their *warm* PP would flow from them into it, which would be experienced by them as a subjective drop in temperature.

The one exception that I mentioned was a single report I read somewhere of a ghost-seeking expedition in which an actual, physical temperature drop of 20 degrees was recorded in the room, but this remains open to doubt. Perhaps the reader may have access to information on this subject. NOTE. This material was writtenseveral months ago and I have since investigated the phenomenon – in a report executed by The Association for the Scientific Study of Anomalous Phenomena, in Belton cemetery where the greatest change in temperature was3 or 4 degrees Fahrenheit, and could easily have been caused by a momentary breeze. It should also be noted that a temperature drop of 10 degrees F. was reported by Long Island Paranormal Investigations. It should be noted, however, that in neither of these temperature checks was an actual ghost appearing to the researchers. The measurements were made at spots which were known to be places where ghosts were known to appear.

It would be instructive here to calculate the necessary transfer of heat from the room to the ghost to see how feasible this scenario would be. The calculation follows:

Let us assume that the room in which six investigators await the ghost's appearance to measure 10 x 5 x 3 metres long, wide and high, giving a volume 'V' of 150 m^3.
Converting this to cc. gives V = 150 x 10^6 cc.
1 'small' calorie = heat required to raise temperature of 1 cc. of water by 1 degree centigrade (or Celsius)
(1 'small' calorie = 10^{-3} 'large' calories. Energy content of food is measured in 'large' calories or kilocalories) The following calculations are in small calories unless otherwise indicated
Therefore amount of heat C_w required to raise 150 x10^6 cc. of water by 20 degrees C is:

C_w = 20 x 150 x 10^6 = 3 x 10^9 calories.

We are, however, working here not with water but psychoplasma, which, in the circumstances being considered, would have a density of about 10 times that of air at 1 atmosphere pressure. Air has a density of 1.29

grams/litre and is the standard against which the relative density of gases is measured. Now we can write:

The density of water $D_w \div$ density of air D_a = 1000 gr./litre ÷ 1.29 gr./litre = 775
Therefore the amount of heat needed to change temp. of air by 20 degrees C. is:

$$3 \times 10^9 / 775 = 3,870,968.00 \text{ calories}$$

And now, having assumed that the PP in the room is 10 times as dense as the air, we can write:
Amount of heat (C_{pp}) required to change temp. of PP by 20 degrees C. = 10 x 3,870,968.00 = 38,709,680.00 calories.
This amount of heat, almost 40 million calories, would be absorbed by the ghost from the PP in the room, if the temperature really did change. I feel practically certain that this invalidates the questionable report of a physical change of temperature. What does happen, I believe, is that PP at body temp. flows from the observers into the ghost leaving the room temp. unchanged. This results, as stated, for the observers, in a subjective experience of a drop in temperature. Let's now compare this with the calorific content of six living human bodies. Assuming that the sum of their Volumes 'V_b' = 0.5 m³ and their density equals that of water, we can write down:

$$V_b = 10^2 \times 10^2 \times 10^2 / 2 = 10^6 / 2 \text{ cc.}$$

and their heat content in calories:

$$`C_b` = V_b \times (\text{body temperature} + 273)$$

(-273 is the temperature at absolute zero where no heat at all is present, which means that the body, at blood temp., has effectively been raised in temp. by 37degrees C. [blood temp.] + 273 from absolute zero). Now we can write:

$C_b = (10^6/2) \times (37 + 273) = (10^6/2) \times 310 = 155 \times 10^6$ calories.

Divide this by 6 to give calorie content 'C_p' of 1 person: $C_p = 25.83r \times 10^6$ calories

Here we see that the supposed loss of heat in the room, if it really happens, would mean that the ghost would have absorbed almost one point five times as much heat from its environment as is present in a living human being. This despite the fact that the definitive ghost is fairly tenuous and would be expected to contain far *less* heat than a human. This suggests, to my relief, that the reported 20 degrees fall in temp. was either false or a mistake, although there is a report in 'Beyond the Occult' by Colin Wilson of an aeroplane pilot or similar airman, who was seen in a perfectly solid, normal state by a colleague two days after he had failed to return from an air-battle, and was presumed dead. However, no drop in temperature was reported. I am fairly sure that the PP density of a ghost can vary between extremely tenuous to approaching a low level of solidity, capable of interacting with matter rather than just passing through it. Writing now a little later, a few months having passed, I see that this could have been a spontaneous and temporary resurrection.

❖ 9. The Conscious Mind is Finite.
The individual conscious mind, I believe, can grow to cover an area of thousands of square metres of the Earth's surface e.g. 100 x 100 = 10,000 square metres, and extend hundreds of metres upwards, but it will be finite and depending on circumstances may have clear boundaries, merge gently with neighbouring minds, or fade into thecollective conscious (diag. 2). The PP of which it is composed will be most dense in the heart and will enable the conscious mind to communicate telepathically with nearby minds or others anywhere on the planet. I have proved this through long range telepathic connections with friends in other parts of the globe, though the relevant contacts were simply *feelings*, and not verbal. It is also a fact that individual minds can differ vastly in size.(See addenda 62)

❖ 10. The biological field (BF)

The biological field, is simply one way of looking at and describing the over ground plus underground material mind, with its tendency to attract or repel. It is, of course, composed of PP and varies in size, strength, form and nature in different individuals. Its form and nature are also very much dependent on circumstances, i.e. the BF of a person walking alone on a dry day in the desert will (I think) be fine, rarefied and approaching in shape a perfectly formed, though changing, modulating, sphere whereas that of someone sitting at home with wife and family will be thick, heavy and indistinguishable from the general mass of PP that unites them. The fact is that the mass of psychoplasma, the BF that is the mind, can take almost limitless shapes and forms and at all times it merges with other minds, if such are present, and the CC and the CU. I remember when I was in hospital some years ago, visiting a new friend on the same ward. I knocked the door of his room, he opened the door and stood there in his underpants. It was strange seeing him like this and I was intensely aware of his, what I used to call 'mental' field, now updated to 'biological field'. The feeling (PP of course) that was emanating from him, and which was extremely strong, warm and physical, so strong that I could 'see' it – with my mind I think, as well as seeing his body with my eyes – and what I saw was a mass of sticky pinkness, (though his body was not really sticky), his torso, as though a bandage had been slowly peeled off a body covered with sticky honey. I myself was in a very delicate, sensitive and anxious state and when I saw his skin, the PP, the feeling, was so strong and *real* that it scared me out of my pants.

I have concluded that BP and spirit are united in the body and are exuded, as PP, through the skin, in all cases, animal or vegetal (BP is made in the cells, PP is a mixture or compound of BP + spirit [S]). Although BP and PP are, respectively, the substances of the unconscious mind and the conscious one, they are so much interwoven in the real world that one can think of PP alone as the main vehicle of the mind. The conscious mind itself originates and is located in the rib cage, neck and head; the unconscious one originates in the gut, pelvis and legs, and is centred at the solar plexus, in the stomach, the home of the god within.

I suspect that my description of the BF is a description only of my own experience of it, and that the average person, not having my knowledge and experience of this subject, would have a much smaller field, perhaps only a few tens of metres in extent. I remember when I lived in Deptford I met a woman near my flat, as I returned from the market half a mile away who told me that she always knew when I was coming home – she could feel it. It varies in nature from person to person and can assume the shape and dimensions of a dome 100 metres high and 500 to 800 metres in circumference, (in my case at least) and at its boundary will become less dense and fade into the surrounding collective conscious. Depending on the subject's current situation it may have at its front an opening 80 metres high which will have the form of an inverted V- and whose PP density will be less than that of the rest of it. (my case again. - in years to come this attempt at exact description may appear as comical to future psi researchers as the leyden jar in early electrical research appears to us, but it is the best I can do at the moment. It fits my observations). It has occurred to me just now that this description of the 'dome of the mind' sorts very badly with my truthful perception of group minds (chap. 7 item 8), and I must leave it, as I suspected, as merely an example of my own expanded BF seen and felt ***from within*** whereas the description in chap.7 is an objective observation seen from outside the phenomenon. Some parts of my description of the 'dome' do make sense. The PP density of this opening or rarefaction which I was describing, and which could exist, may be no more than the basal density of 0.001 which I assume to be the lowest PP density to which the CC can fall, while that of the main part of the dome, could be four or five times this amount. This is because the subject can directly see other people, friends or potential enemies in front, whereas the field at his back will sense their presence behind him, which tallies nicely with the next item (item 11- making a distant person turn round etc,). Its size, form and nature will depend on the position and current activity of its possessor, e.g. the detailed description just now given will fit the case of someone standing in the park being made to turn round by an observer. These particular, specific characteristics of my BF were largely a consequence of my extremely insecure state of being at that time. Things are much more relaxed and 'normal' these days.

It seems very likely that PP is full of information in some state which is how and why thoughts and feelingsare transferred from one mind to another. Some people have a light, airy and almost imperceptible field, while others have a thick, heavy, enveloping one, but the biological field of almost any person will vary according to circumstances. The thick, turgid BF which some people possess, can throw a spell over anybody in its vicinity so that the person so afflicted has to get out of the room as soon as possible. Such a field will not affect everybody in the same way. Some people, especially friends of its owner to whom it will be a reassuring feeling of friendship, will not be sensitive to this overpowering biological phenomenon. I think that the BF is naturally and instinctively drawn in by its owner like a large, warm cloak, as he enters a building or smaller space such as a room, although he remains in contact with the collective conscious and the collective unconscious which permeate and pervade the walls, ceiling and floor of the room.

NOTE. I am beginning to suspect that these differing shapes, forms and fields are all the same thing – that is –the active conscious mind, which can take any shape or size necessary to engage with its current activities, but even so I think it better to leave them in the essay since it is often useful to have something described from more than one point of view; also it is often helpful to be shown the steps by which a conclusion was attained. I am sure that the conscious mind is finite, and the unconscious mind is not.

❖ 11. Making a distant person turn round by staring at his back
For many years now it has been a puzzle as to how one can make a distant person turn round by staring at his back. When I was in hospital some years ago and well-advanced along my path of discovery, I noticed that a female patient who habitually kept up an unbroken stream of verbal gibberish, always stopped abruptly whenever I thought certain thoughts. Precisely what thoughts I don't remember. There was also another lady who succumbed to a dramatic and sustained burst of flatulence on similar occasions. These amusing events and similar ones lent an additional and welcome confirmation of my my concept of 'close-quarters telepathy'. It happened with other patients from time to time, and was an impressive demonstration of one body influencing another through a somato-psychic-psycho-somatic causal chain.

It is a fact known by many that in conversation in the home, workplace or pub some mentally powerful people are able to manipulate the mind, thoughts and behaviour of others. An illustration that springs to mind is the cruel Don Juan figure who retains an emotional grip on a lover who wishes to be released, knowing that he no longer loves her. He does this purely to gratify his lust for power and control. In the home or the pub the PP would be fairly dense, but in the park assuming this to be the location of an experiment in making a person turn round, it would be present in various densities depending on circumstances.(See fig.12).

This, obviously, is what happens when I stare at someone's back in the park.They are 100 to 200 metres away. I am staring at them which causes a gradually increasing PP connection between them and me via their BF the CC and my own BF which *feel* the attention. We would already be connected on an unconscious level by the universal unconscious. Then they turn to find out who or what it is that is disturbing them (see 'group minds' in chap. 7 item 8).

Going back to my experiences in hospital with the verbose and flatulent women; on thinking further it seems pretty likely that the thoughts in my mind which caused the phenomena, and which I think they themselves might have been encountering in their own minds as a result of telepathy via the plethora of PP in the room, created thoughts and feelings, which they tried to comprehend. In the case of the flatulent woman this would entail a high degree of concentration at head level which would leave the gut 'unattended' so that it would weaken thus allowing an escape of 'wind'. The other, hyper-loquacious woman, I think must have experienced a sudden feeling of peace or connection with someone – me – which caused her to cease from her purposeless verbal meanderings. If she had been a different sort of woman, someone to whom I could relate, I would have made friends with her and lead her out of her inner chaos.

❖ 12. The non-distance-related nature of psi
In electromagnetic communication there is a distance-related characteristic, that is, the signal becomes weaker with distance, which is due to its radiation-based nature and is described by the inverse square law. This states that the strength of a signal decreases in proportion to its

strength at its source divided by the square of the distance between source and receiver. In distinction from this, psi phenomena are conducted through space by a medium, psychoplasma, the density of which varies between 1.00 which is the density in the heart of a healthy subject and 0.001 which is my estimate of the residual or 'basal density', below which the density does not fall wherever it is measured. (There are occasions I believe, when the transmission will be instantaneous, see chap.5 item 13) The density will depend on location and other factors. That is, a measurement taken on the open road between London and Dover will be less than that within the confines of either. The basal density may be considerably less than the density of the atmosphere at sea-level.

> This, I think, is why psi is not reduced by distance. The basal density is sufficient for good contact. (Graph 1 and graph 3). The mechanics of how all this happens is, by virtue of the fact that between a telepathic connection's origin and its destination, no matter what the distance, the density of the collective conscious will vary according to location and circumstances, including poles and deserts where D = 0.001 (basal density). The PP density will rise steeply in populated regions, as shown in graph 3. (see chap.7 items 7 and 8).

The means by which the PP density is maintained apart from that bequeathed by generations of evolution, is the presence all over the globe of numberless millions of living organisms, flora and fauna, all of which are constantly generating psychoplasma. Any single, larger organism, as already described, will contain PP at D =1 in the heart, and the density of its PP would, in theory, decrease, to $D = 1/\infty$ with increase of distance, but, as shown in the previous sentence, the general PP density will be maintained, so that the density of the CC will remain at or above its basal density all over the globe. In addition to this the PP of these PP-generating organisms combines with loose, unattached PP, to help form the collective conscious. If, on measurement with the suitably calibrated vacuum wheel PP detector, it should be found that PP density really does vary by significant amounts, a unit of measurement will be needed; I think 'Rhine' would be a very suitable candidate for this function, since J. B. Rhine was the first one to put the study of psi on a firm, scientific footing. (Graph 3) and also had the courage and integrity to reveal the presence of something in the human being which could be described as the soul or spirit.(see ref.10)

❖ 13. The Einstein/Podolsky/Rosen paradox..

In 1935 while based at the Institute for Advanced Study at Princeton, Einstein and his two younger associates constructed an apparent paradox, a 'thought experiment', the result of which came to be described, by Einstein himself, as 'spooky action at a distance'. It depended, for its effect, on the fact that in quantum theory, if two systems interact they become permanently connected and any change in the state of system A causes a parallel ***and instantaneous*** change in system B no matter how far apart they may be. This is in direct contradiction of the theory of relativity which claims that information sent from a given point in 4-dimensional spacetime cannot travel faster than light, which limits its speed to 186,000 miles per second. This means that it must take a finite period of time for information to make the journey.

The actual imaginary systems they used were two electrons which were in a 'singlet' state where their spins cancelled out to give a total spin of zero, and, in imagination, they drew the electrons apart and then measured the spin of one of them. Since, in quantum theory, according to a resolution known as the 'Copenhagen Interpretation' in which Niels Bohr and others agreed that the state of an atomic system was undefined until it was measured, this measurement would actually have been given a value by the very act of measuring it. Then, in order to preserve the essential total spin of zero, the other electron would have had to *instantly* adopt the opposite spin, whatever its distance from the first electron. Einstein saw this as a demonstration of the incompleteness of quantum theory, believing, as did everybody, that nature is 'local'. This locality of nature means simply that if an event occurs at point A in space there must be an interval of elapsed time before its effects become apparent at a distant point B.

This is a natural consequence of the effects of the Theory of Relativity on space, time and the universal speed limit, the velocity of light.

This is why distant stars are seen not as they exist now but as they were several years ago – their light – a physical thing - has taken all this time to reach us. This too is an instance of the locality of nature, which everyone, including Niels Bohr and his fellow quantum theorists, believed to be true. What Bohr, Einstein's intellectual opponent, condemned as impossible was the assumption that two electrons with opposite spins forming a 'singlet' could be separated. – pulled apart, which would imply that each of the electrons would be existing in a local or separated state, an apparently naïve suggestion for in atomic systems according to quantum theory, once atomic systems have been connected they can never be separated.. Einstein, who had reservations about the validity of quantum physics, believed that the value of any quantity (a) separated from a quantity (b) for some time could not be affected by measurements made on the latter, which, if true, meant that locality prevailed.

Thus, Einstein and his friends thought that they had demonstrated, in their 'thought experiment', the existence of hidden variables or elements of reality, (e.g..evidence which might be discovered at a later date, thus making quantum theory complete and less mysterious) thus showing quantum theory to be incomplete, but they had no experimental evidence, and did not seriously believe in non-locality anyway. Therefore the bizarre situationremained unresolved; relativity which was unassailable was in conflict with the 'Copenhagen Interpretation' which was, at the time, equally unassailable. This state of affairs endured for thirty years but then a Belfast physicist John S. Bell on a one year leave from CERN, the European Centre for Nuclear Research, developed an 'inequality principle' to test the questions raised by the paradox. To derive his inequality Bell used certain facts and ideas with which everyone could agree, except for…… Einstein's condition of locality, which he assumed to be true. If now it t6666666t6were shown by experiment that Bell's inequality principle could be violated this would mean that one of the premises in his derivation was false. He chose to interpret this as meaning that nature was non-local.

Some years later experiments by John Clauser at Berkeley in 1978 and evidence from experiments done in Paris by Alain Aspect and his group, in 1982, indicated that 'non-locality', a necessary consequence of the rectitude of the EPR supposition was, in fact, just that – a fact. Quantum entanglement, or 'spooky action at a distance' was indicated as experimentally verifiable. This means that despite the local appearances of phenomena, our world is actually supported by an invisible reality which is unmediated and allows communication faster than light, even instantaneously. This may be a contradiction of the theory of relativity but it is not a contradiction of the power of the god within, operating in a metaphysical universe, to travel, instantly, possibly at times transporting matter, from anywhere and anywhen to any otherwhere and otherwhen. (see chap. 4 item 11). Taking note of all this it can be seen that this attempt by Einstein and his friends, to expose quantum theory as incomplete, backfired and ended up finally by showing that Nature was 'non-local' after all, so that faster than light transmission of information is possible, a fact which the theory of material mind informed by the god within tacitly allows anyway, though no instances of this have been observed and no attempt to measure the speed of paranormal communication has yet been made to my knowledge. Thus physics has played into the hands of parapsychology, my version of which itself proclaims that total, inclusive knowledge of Nature and her ways, will never be attained by science anyway, nor by any other human institution for that matter. There will for ever be the eternal mystery of the gods within, and presumably without, and the only way that humankind can gain complete and perfect knowledge is by becoming as gods ourselves.

CHAPTER 6

Love and psychoplasmatic bonding

1. The sexuality of the god within; 2. Falling in love; 3. Omni-love and self-love; 4. Time, space and the afterlife; 5 . The universal conscious; 6. Precognition; 7. Psychometry or token-object reading.

❖ **1. The Sexuality of the God Within**

As stated earlier the god within is omni-sexual but its activities and appetites will be determined by the spirit in the heart (the essential self), and the head working in tandem, which together make up the conscious mind, which is where the sexual orientation of the individual, hetero- homo- or bisexual is located. Instinct is unconsciously motivated and is therefore based in the gut.

❖ **2. Falling in love**

There is a love-song that goes:

> 'So please forgive, this helpless haze I'm in,
> I've never really been,
> in love before,
> and now it's only you, it's you for evermore'.

The 'helpless haze' is the result of two minds, material and made of PP, mixing so that the mind expressing itself in the song is no longer alone, independent and free, but is flooded with a confusing mass of new, unknown material in the form of PP, some of which supports known beliefs and some which contradicts them. This would apply particularly in the case of a young and inexperienced person whose beliefs, opinions and value systems were not yet very developed. An older, stronger mind would keep a firm grip on itself; the body's long-term memory centres, heart and gut, would be well-filled and therefore armed against the seductive forces of the other's PP. I think everyone must have felt, at some time or another, the warm, attractive 'charge', the 'oomph' when one gets near to a member of the opposite sex, especially if that member is young. This is, of course, a function of the density of PP present, and it isn't necessarily mutual, although of course it can be.

❖ 3. Omni-love and self-love

We are interdependent entities. We are all joined at the gut, the home of the GW. I love my wife and male friends and am happy to tell them so. They are all joined to me at the gut, though my wife is not, or only weakly, joined to them, being strongly joined instead to her own friends. I like my female - probably platonic – friends though we are not strongly joined at the gut. I love my dog or cat to whom I am strongly joined, and we shall have grown to be part of each other. Like Beethoven, I love the human race despite its multifarious foibles, cruelties and inadequacies. My male friends and I are 'attached' to each other but not in a figurative sense; the connection is real, material and gut-to-gut, namely, bioplasmatic, the medium of the GW. Thus it can be seen that any particular GW, in its home in the PU of a specific person or creature, is generous and retains its essentially good and enthusiastic nature irrespective of the person or animal to whom it belongs. For these reasons it can be described as androgynous, the deployer and generator of 'omni-love' – a love that binds all life together.

The private and personal self, however, is to be found in the heart, and its attitude, mood, or state of being can change radically from moment to moment or day to day. It is always in closer or more distant connection withthe GW and at positive, happy, times the co-operation of the GW and S can increase dramatically so that the heart is filled with enthusiasm, but at other times, depending on the attitude of the subject, i.e. negatively rather than positively disposed, the heart may be filled with despair.

❖ 4. Time, space and the afterlife.(See addenda 42)

Time, as we know it I think, doesn't seem to exist in the afterlife, where any one of the departed is free to visit any other, also departed, a friend or relative perhaps, since both of them are 'in spirit' and 'outside' time; it is also able to visit any other place on Earth while staying in the Afterlife, and any time in the past. The afterlife is certainly accessible through telepathy and one of the more astonishing conclusions to be drawn from this, and the fact that everything that lives or ever has lived, is joined by the BPE, as written earlier, is that it is theoretically possible to resurrect (teleport) one or more items of any organism, animal or vegetal (see chap.8 item 6 no.9 no.24) from its place in the AL to the PL here and now. I do not think it would be possible to do the same for the future, since the GW might object, i.e. one could see one's own death, or resurrect one's friends in a state of old age.

It seems that the human race is experiencing a profound and highly significant evolutionary advance. A mental and spiritual one. It would seem that in the afterlife, things that would normally occur with decades between them can all happen on the same day, (see contacts with Leonardo in 'author's experiences'chap. 8 item 6) and, by inference, in, on or around a timeless and spaceless region. I can infer from this that everything that ever lived but is now dead is now in the afterlife, which exists in, at or around this region. This, if true, offers a solution to the conundrum of where do all the creatures from the past, men, women, hominids, dinosaurs, sea-creatures, forests, gowhen they die? They must go to the timeless, spaceless region suggested a few lines ago. If this were not so, and they went to inhabit some other region of space as conceived by us, a very rough estimation of the area needed to accommodate them would be the number of generations since life began - a very rough figure – times the area needed to accommodate one generation. The total area needed to accommodate all current life on the planet cannot exceed 200 million square miles since this is the planet's approximate surface area capable of supporting life.

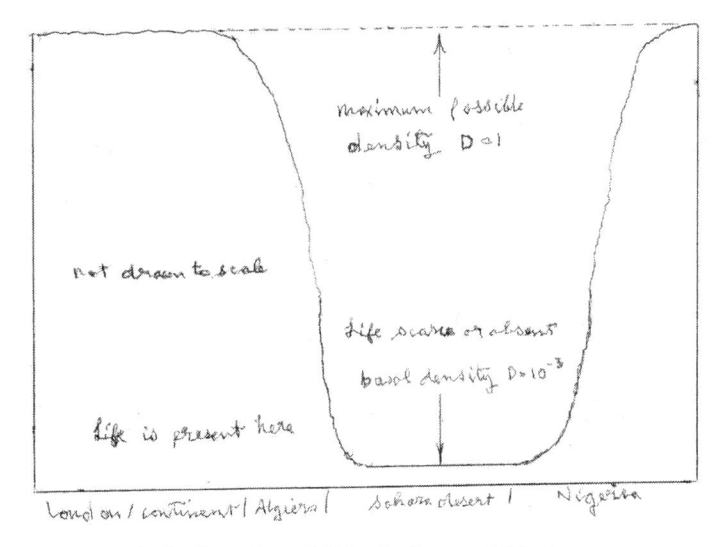

[Hand-drawn graph showing density curve with annotations:
- "maximum possible density D=1" (at top)
- "not drawn to scale"
- "life scarce or absent basal density D=10⁻³"
- "life is present here"
- x-axis labels: "London / continent / Algiers / Sahara desert / Nigeria"]

Graph 3. Density of PP relative to differing parts of the landscape and flora and fauna.

Assuming that the larger life-forms arrived on the scene about 1 billion years ago, and further, that a generation is about 10 years, (an estimation of the average generation of the combined differing generations of all kinds of living forms, from mayfly to californian redwood tree), the total area needed to accommodate all life that has ever lived is $2 \times 10^8 \times 10^8$ which equals 2 quadrillion square miles, a very rough but significant figure, equivalent in fact to 100,000,000 planets the same size as the earth. When I say all life on the planet, I mean literally all. People, animals, octopuses, forests, mosses, insects, marine life, birds, whales, micro-organisms, plankton, etc.. They are all joined by the BPE and all go to the afterlife. Of this I am certain.

At this point in my essay I must reveal that in the last two days a completely different, and much more cogent, vision or model of the afterlife has surfaced in my mind. Basically I see now that the departed being is not reincarnated but exists as a psychoplasmatic presence which can be resurrected, i.e. teleported, to the here and now. This PP presence is non other than the conscious mind plus the unconscious mind, (spirit plus god within) of the departed organism, both of them material due to the presence of BP. It may be objected that such creatures as mice and ants do not have minds but I would disagree. I believe that every living organism, even down to bacteria, has a metaphysical component that goes to the afterlife and which can be resurrected or materialised although this may be difficult with such things as californian redwood trees and dinosaurs, of course, because of their size, and unwise in the case of the smallpox virus! The only respect in which the departed differ from the living is that they don't have bodies, and therefore cannot make their presence known to us, (omitting Plato's angry attack on myself chap.8 item 6 no.60, and Heraclitus and Pythagoras, which tends to invalidate this observation) except through the efforts of mediums, who, presumably, use telepathy and some kind of spiritual vision. This fact would also suggest that the departed have no access to their memories since these are stored as traces or feedback systems at various locations in the body, a possibility which disturbs me.

There is, however, another aspect to the mechanism(s) of memory. Sometimes, when one 'casts one's mind back' to locate a memory of the last few days or longer, one mentally reaches into the PP to access a region which lies behind one's head at about six to ten feet above ground level. One also looks heavenwards to seek inspiration and /or access memories or information. This suggests that there exists, in addition to memory centres in the body, a non-corporeal memory system or region which is presumably symmetrical to the corporeal one, and which contains the same or additional memories. This would seem quite cogent in view of the fact that we have concluded that PP is stocked with information as a necessary explanation of the mechanism of telepathy. This apparent problem with memory disappears if one contemplates the fact cited (Chap. 1 item 5) of the late Kenny Everett summoning the most ridiculous and amusing ideas possible from his fertile imagination, and my own staged ribald 'insulting- matches' with friends in which each tried to outdo the other in his outrageous and comical 'insults'. As written I remember the sensation of creating one's insults as a feeling of reaching up with an imaginary hand, into the space two feet above and behind one's head, to grab a bunch of crazy ideas. I see this very clearly now as reaching up – literally - to grab, with the imaginary hand, a handful of PP which would be full of these crazy ideas – in other words – a mass of information which would, of course, be memories as well as ongoing psychoplasmatic activity.

I think, in view of this, that it is possible that the memories lodged in the body, especially its lower parts, are probably permanent while those outside the body in the surrounding PP are in a state of flux. This reasoning also lends weight to, in fact agrees with, the idea that PP is saturated with information, something which, I think, is conclusively demonstrated by 'the PP-grabbing hand' mechanism. This discovery multiplies the opportunities of investigating memory, and these can be found under 'Memory' in Chapter 3.

This new vision also neatly explains the conundrum of locating the afterlife, and the fact that I didn't see any surrounding furniture or environment when I made my contacts. As stated, I think that PP must be saturated with information in some form or other, perhaps a subatomic storage system (See ref. 7), in such a way as to make it possible for many beings to occupy the same space, that is, any given volume of space; i.e. the volume of an average room in a house, could be occupied by several human beings, lots of animals, trees, plants etc., all merging into one another, in some way, so that a space which in the current life could contain only six human beings, a few dogs and cats, plants etc. would be able to hold several hundred humans, two hundred dogs, cats, etc. and ten gardens full of vegetation, or even a thousand times this figure. Extending this idea (in view of the estimated space required for re-incarnation being equal to 100,000,000 Earths) it may be that a given organism in the AL occupies only a thousandth of what it occupies in the PL – or a millionth – so that the two metaphysical entities the GW and the spirit plus the associated psychoplasma, which would in any case be part of the collective conscious wouldn't *need* much space, if any, in which to exist – they are after all quasi-metaphysical entities, which would mean that the billions of cubic kilometres of the psychosphere could possibly accommodate an infinitude of departed souls. Add to this the likely fact that any given organism probably remains at or around the place and point in time at which it dies. This latter supposes that the afterlife may be layered around Earth like the layers of an onion all through time, to include every geological period since life began, which itself reduces the problem of accommodation of the billions of life forms which must have existed.

To return to our discussion of time, this very day, walking to my studio, I noticed one of the small, cast-iron covers that the water board puts in the pavement. The word 'WATER' was printed across it in raised letters and I immediately thought, "those Sumerians who invented the cuneiform script 5,000 years ago had no idea that a development of their letter-system would be on this metal tablet 5,000 years in the future", and it suddenly felt as though I was with them watching them writing and simultaneously looking at the cover. I was persuaded by this that the past is still with us (it can be remembered – and visited) as is the present but the future is unreal or at best is happening right now, with the advancing moment, which leads to the question of precognition –a well attested phenomenon. This, of course, is not counting recurring cyclical events such as sunrise and the seasons.

Diagram 13. is a worldline for four events, the space below the horizontal dividing line represents the past which is real and can be explored, that above the line is the future which is unreal not having happened yet, and the line itself is the present moment, the 'now'. A and B is a race between two horses the winner of which is seen in a dream; C is two ships on a collision course whose pilots have miscalculated; this too could be a 'premonition'. D is another precognition seen in a dream. The position of the pyramids will, of course, not change with the passage of time. An explanation of the worldline is given near the end of the addenda (no.120)

I felt that they formed an unchanging plenum, and I was led to conclude, as I had before, that there is no such thing as time, but only change and information which is distributed throughout the body and the surrounding PP, and is a record of all its experience, both internal, imaginative and material received via the senses. This leads us to speculate on the subjective nature of 'personal' time, that is, when one is happy time passes quickly and the reverse if one is unhappy. Subjective time is just a feeling, like boredom, excitement, joy, misery all of which have a 'time-related' component.

There is a conception of the nature of time known as 'the tenseless theory'; It exists in a 'block' universe where, in theory, everything that has existed, is existing, or will exist, is real and in existence at this moment. I myself used to believe in this theory but no longer do, feeling that time has no real existence and have put in its place the chinese concept of change.

I read recently that Einstein and his friend Kurt Gödel, the mathematician, were talking one day and Gödel informed Einstein that he had extended the Theory of Relativity to show that time was not only related to the velocity of the object, but that there was no such thing as actual measurable time. It seems certain that time does not exist in the afterlife. I know that it is possible to travel into the past from one's position in the PL by means of a temporal teleportation, having proved this for myself with a thought experiment in which my head and its surrounding PP were actually partly in the AL by virtue of the materiality of the mind (PP of course). A thought expt. can use established facts to deduce or predict resulting ones. Travel forwards in the AL may be possible only as the result of a teleport. by some future person, as was the case with Pythagoras, Plato, Heraclitus and the hominid who arrived as a result of my thinking of them; this itself suggests that there are, or rather, will be, people in the future which is obviously true.

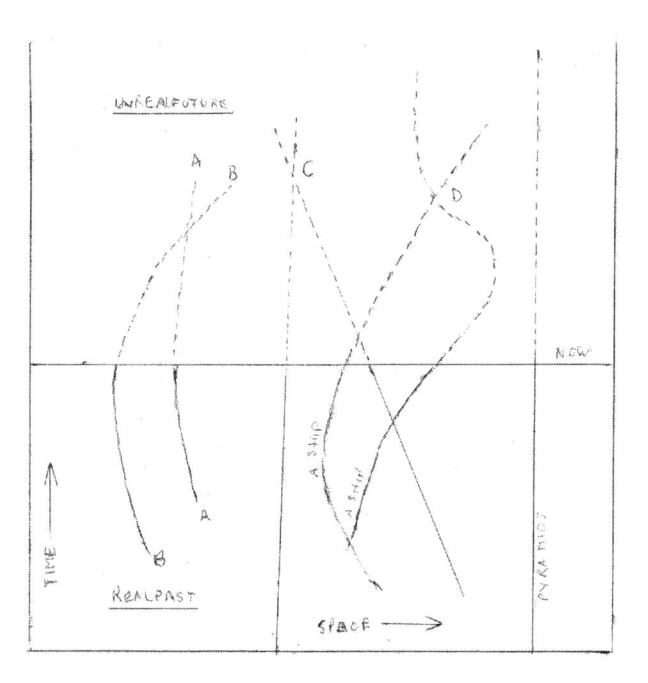

Diag. 13. World line for two ships on a collision course C, a horse race A and B, winner seen in dream, and D two other ships and a premonition and the pyramids whose spatial position does not change. (Explanation of worldlines shown in Addenda 120).

The concept 'time', I believe, is simply a convenient and natural way of keeping track of change. It is a mental device and has no objective reality. In South America there exists a people who have no words which relate to, or signify, the concept of time. Their way of indicating time or the succession of events is to make appropriate bodily movements and gestures. If one of these people wants to refer to the past he simply points back over his shoulder. The redskin talks of 'many moons' or 'many summers'. These naturally occurring cyclical events start with the period of the earth's rotation, the day. Longer periods are represented by the phases of the moon, vernal and autumnal equinoxes, summer and winter solstices and finally the year, the period between winter solstices. The only real, observable, external indication of 'time' is change - of state, temperature, colour, composition, position, of a shadow thrown by the sun, etc., just as the only real, observable, internal indication is feeling, or one's state of being.

As far as measuring time is concerned, if analysed, any system designed for this purpose will be found to be a recurring change, e.g. the hour-glass, or the Egyptian 'candle clock' or a regular repeating cycle, e.g. the pendulum, the balance wheel of the clock, or the period of the caesium atom. The balance wheel of the clock oscillates 'x' times during the day, whose duration has been determined presumably by convention. The scientific definition of the second has been set as 9,192,631,770 vibrations of the caesium atom and the atomic clock in Boulder, Colorado known as NIST F-1, and a similar one in Paris - 'atomic' clocks which use this system - are the most accurate clocks in the world today, and will lose less than 1 second in 20 million years. Every one of these clocks, however, simply takes some object or mechanism which seems to be cyclical and sets it as the basic unit of 'time' (or change). Galileo used his pulse rate to measure the period of the pendulum in the cathedral.

Using the tensed theory of time and the mechanism of teleportation which I know is possible, it should be possible to teleport a person, organism or a piece of inanimate matter, from any one place in spacetime to any other.

(Diag.3). I see this as becoming possible within the next 100 years. Diag. 4 illustrates an occasion when I myself almost did a 'timeflip' back to age 14. I was sitting in my room brooding on this or that when I suddenly caught myself looking down and into the past, at an image of myself setting up the aerial of my 'crystal set', amateur radio being my hobby at that time. The strange thing was that as I looked I became aware that I could do a 'flip', a sort of 'forward roll' or somersault into the 14 years old self that I was looking at. The only thing stopping me was fear of the possible consequences, and I quickly pulled myself away, shook myself free of the vision, or feeling, for so I could call it.(It possessed both of these qualities). I am quite certain that I could have done the flip since the feeling I had at the time was exactly similar to the feeling of holding back from 'the rubicon', or a sexual orgasm, when it is vital to terminate proceedings having forgotten to take contraceptive measures.

If I **had** done the flip I'm pretty sure that I would have found everything exactly as it was in 1952 when I was fourteen. I would have gone home to the same, younger mother, same house, same brother and sisters, same fifties adverts on the hoardings and the walls of the same public houses, and parked outside the same fifties-style cars. I would then, I am sure, have had to grow all over again to arrive back in this room where I am typing. It could be argued that this idea conflicts with the principle of freewill, since, if I grew back to this room and this PC, I would have exactly duplicated the choice of actions that got me here in the first place; however, thinking about it I feel that that is exactly what would have happened; my choices would have been determined; this, at first sight, is not the time travel talked of in physics where one, presumably, keeps one's present form and clothing and does not become a younger or older person. In passing I should mention that I have learned that this feeling of being on the edge of a parapsychological event (e.g. a timeflip or a teleportation) as just now described, means exactly that; one *is* on the edge of a psi event. This subject is dealt with in greater detail in another part of the essay.

It is interesting to speculate, assuming that I had done the 'flip', on how I would have felt at the moment of return. I think it is likely that I would have arrived back in my room to find myself pondering on the advisability of doing the flip. This again raises the question of freewill. Would I have done the timeflip again? This, I think, depends on whether I was aware that I had just then done it. If I **was** aware, it would obviously be useless to do it again (except for revisiting good times) so I would probably continue into my future. If I wasn't aware the horrible possibility of a recurring, cyclical timeflip would arise; I would have been 'trapped in time'. This possibility, plus other considerations e.g..the flip would then have been added to my (presumably remembered) experience, suggests that I would know that I had done the flip.

It also occurred to me to compare a spatial teleportation with a temporal one and it seems that in both cases the essential mechanism is that matter is transformed into PP (in the case of living things) in which form it can be transferred from place to place in space or time. (I am not sure whether it is possible to transform inanimate matter into PP, but I know that it can be teleported). I am sure, actually, thinking again, that it is possible since if it couldn't it would mean that a teleportee would arrive at his/her destination bald and naked if one assumes that human hair is dead matter, and I'm pretty sure that such occurrences have not been reported. The spontaneous teleportations that I, myself, have witnessed have all been associated with a physical body, wearing clothes and certainly not bald. Finally, it is possible that an inanimate body can simply disappear from spot A to instantly reappear at spot B by virtue of the mysterious god within; this interpretation would perfectly fit the case of the guitar, spoon and canister of body-spray which arrived in my room a few months ago. As far as I know it is posssible that they disappeared from one spot to reappear ***instantly*** at the other thus exemplifying the anti-relativistic effects exhibited in the EPR paradox (See Chap. 5 item 13) To continue; the spirit that I was then, plus its attendant PP, if I had allowed the 'time-flip' to happen, would have 'migrated' through the unbroken mass of PP that stretched back through the days and nights to arrive unchanged at age fourteen, except for the fact that it would be smaller, since my sojourn on Earth at that point, being shorter, would not have given my conscious mind time to grow to its present size . It should be noted here that the spirit and identity of any person 'A', remains inviolate as 'A' and nobody else. There is only one of this particular A and it retains its identity throughout whatever temporal or geographical teleportations it encounters.This is an example of the 'Individuality of the Spirit' (IOTS).

So far, as already indicated, I have constructed two possible scenarios for the afterlife:

Scenario (1): Reincarnation, i.e. real people with flesh and blood bodies breathing air and walking on the ground, or rather, *a* ground, but living in a condition of timelessness and spacelessness. Obviously wrong I would think, although I did see T. S. Eliot raise his weary head. I was going to write just now that I'd contacted my two cats but as the thought occurred I suddenly became aware of a greyness and claws in the air above me. I had to stop. I also connected with a 7,000,000 years old hominid to the extent of almost resurrecting it, and had to shake myself clear of it. If one thinks about it, my image of T.S. Eliot 'raising his weary head', is not necessarily an indication that the departed have flesh and blood bodies, but merely a temporary, partial materialization which was not allowed to continue to being a full resurrection (or materialisation)

Incidentally, and this is not really pertinent to the theory but merely a space–filler, (I have lost a few lines of text) It occurred to me a few weeks ago that the perpetrators or martys of 'suicide bombs'could, with the technology of TMM, be resurrected to proclaim their disappointment and anger at finding no sign of the seven virgins that they were promised as a reward for their self sacrifice. That would put an end to the nonsense.

Scenario (2): When psychic mediums talk about their activities they refer to the contacted person as being 'in spirit'. I have noticed that on the occasions when I've seen a medium in action (on TV) they sometimes answer the client's question "where is my sister/brother etc. now?" by saying, "he's standing behind you" or "he's standing just there, in front of us." Discounting the fact that they could all be charlatans, and many indubitably are, this suggests that the spirit, plus presumably, the GW, really is in a state of disembodiment and is merged with a host of other spirits, so much so, that the billions of organisms that would require 20 quadrillion square miles in scenario 1, could, in the latter one be in a state of intermingling or loose coalescence, occupying a space of the order of seven or eight kilometres deep which surrounds the globe as the peel of an orange surrounds its contents, so that they could be summoned by name just as one summons a site or person on the internet. If this were indeed the case, it would still be perfectly easy, in theory, to resurrect organisms, and it would still be possible to see T.S. Eliot raise his weary head, almost see my parents' faces, and feel Leonardo's presence and personality.

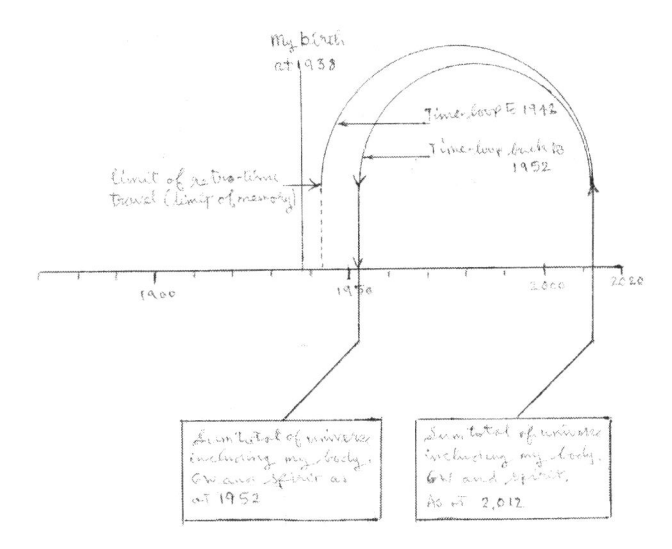

Diag. 4. Retro-time –travel through
psychoplasma = temporal teleportation

There is one small problem with this idea and that is, unless one assumes that the GW and the spirit, as metaphysical entities, occupy either a very small space (or no space at all), the projected psychosphere may not be capable of accommodating them. I mentioned this matter a few pages ago and the positive conclusion, quitepossible, was that the psychosphere may afford space for an infinite number of metaphysical beings.

How does this stand up to the actuality of the BPE? The BPE connects the afterlife with the present one; this is beyond dispute, and is in fact the means by which the AL was discovered. The active principle in the BPE is the god within, that is, the billions of divinities which amass and deploy BP in the universal unconscious. (the UU is identical with the BPE). This being true we must assume that any given discarnate spirit in the AL must be accompanied by its GW or attendant divinity. There is also the curious business of seeing Leonardo as an older man and as a young one in the space of a single day. In both of these contacts I felt his mood and personality. After some consideration I see – and realise – that this happened simply because I possessed ***two*** pictures of him, and since my parapsychological researches have given me a great deal of power to help or hinder others, when I summoned the younger picture to mind the relationship between him and myself must have been very positive leading almost to a resurrection or teleportation, in other words, I made him ***feel alive.*** When I summoned the older picture things were very different, I'm pretty sure I made him feel old, as he actually looked in the picture, which would not please him, of course, and it is possible considering the following, that in doing so I was actually dragging him away from a younger conception of himself, to an unwelcome state of age; that is, it is possible that the organism could exist at any preferred (by it) stage i.e. when one has died one could then choose a period from one's past when one was happiest, and elect to stay at that age, or to ***feel*** as one did then, no matter what one's chronological age, which would obviously be greater. (This brings to mind my own near 'timeflip'to age 14, diagram 4. p. 124.)

My second conception of the afterlife would require less room for its existence than my first one since no physical bodies would be present. Assuming that it would require the total volume of the psychosphere (PS) for its accommodation, and that the PS would rise to a height of eight kilometres, neglecting its underground part, we can calculate that the available volume would be 4.1263 billion cubic kilometres. In this scenario I see departed individuals, human, animal and vegetal, that is their spirits and attendant divinities, as dwelling not only at ground level, but being capable of floating, drifting or rapidly moving anywhere, even through each other, in the space enclosed by the psychosphere. We must not forget, however, that the AL and PL share the same space. There is one difference though – departed souls can exist literally anywhere and I suspect anywhen (except the future for reasons given), the psychosphere – even eight kilometres above ground level, whereas presently living organisms are confined by gravity to fairly near ground level.

After some rethinking I have developed the following picture of the afterlife, which is essentially my second conception clarified and reorganised. Unfortunately it may duplicate some ideas already expressed in other parts of the essay, for which I ask the reader's forgiveness.

- The afterlife (AL) and present life (PL) share the collective conscious (CC) but in this revised scenario it may be called the Universal Conscious (UC).

- The collective unconscious (CU) of the PL is amalgamated with the CU of the AL to make the UniversalUnconscious and the UU = BPE which is the most powerful entity in the psychosphere.
- Hypothesis: the average density of the BPE grew slowly throughout the geological ages, that is starting around the Devonian age (the age of fishes) through the various ages, carboniferous, Permian, Jurassic etc. to the present day. That is, the at first extremely tenuous BP which would have accompanied life on Earth (bacteria only) before the dawn of the Cambrian and Devonian periods, would have reached a considerable average density by then, although there could have been 'holes', in deserts for example, where there was no life and, consequently no BP. As the density continued to grow, however, these holes would have been flooded with tenuous BP so that finally there was no place on Earth that wasn't included in the BPE.
- The density of PP never falls below a basal level which I estimate at 0.001 of the density in the heart which it is convenient to set at 1.00, but increases to a maximum of 1.00 whenever and wherever several bodies congregate, indoors or out.
- My current picture of the UC is of a clear, still or moving volume of fairly tenuous PP which covers thesurface of the planet like a blanket and varies in density from place to place.
- Stating the already stated but in a clearer form: the PL and AL share the Universal Conscious (UC) and PL and AL share the Universal Unconscious (UU) = (BPE). If we think back, or rather forward, to 'group minds' (chap. 7 item 8) we are granted a more empirical perception of a 'group mind'. On that occasion I was directly aware of a large blob or mass of PP which surrounded and enveloped a family and their dog. The PP could not have been more than five or six times more dense than the air in which it floated, but nevertheless I was very aware, later in my room, as I wrote of the incident, of a transient connection with the man in the family and his displeasure at being reminded of my perverse failure to join his little group. (See graph 1 and Chap. 2 item 5 a). This later connection with him is further evidence for the existence of a PP cut-off or 'Basal Density'.

A thought has occurred to me. If PP passes through solid matter, how does it cause PK? The answer is: only the spiritual component of PP passes through the object – PK is caused by the material BP component. This is more food for thought. I notice as I try to imagine or conceptualize what is happening here, that it feels as if I am developing and strengthening my limited PK powers. I also notice, in this respect, that I have developed the power of seeing 'group minds' as just now written. See also Chap. 7 items 8. 9.

There also remains the need for an explanation of the fact that I never saw the circumambient environment of my AL contacts, or their faces or bodies except as represented in pictures that I possess. I would see their faces and bodies only as the result of a partial materialization (teleportation). This, actually, if my second picture of the AL is accepted as true, as I believe, is no longer a problem because the environment of the departed being would be the CC like my own environment, seen all around me. In the case of a connection with Leonardo, say, who would probably remain close to his previous life's vicinities, I would be in contact with the CC at my computer, which would merge with the PP of his mind – him, via the CC – in Florence or Paris where he used to work. I would be in telepathic contact, getting his current feelings and mood (slightly risky getting so close, for fear of an accidental teleportation, amounting to a resurrection) but seeing him only in terms of one of the two portraits of him which I have.

NOTE. I have just now had an inspiration. I had earlier concluded that a slightly depressing aspect of the AL, is that, although one would certainly rejoin one's family and friends, one would not be able to see them since they would not have bodies; but add to this the fact of my own growing powers of seeing 'group minds' (chap. 7 item 8) and it would seem likely that the departed probably are, or at least can be, capable of seeing each other in the way that you and I see ghosts – a faculty which could grow stronger with the passage of time. They are certainly capable of seeing the real environment in which you and I live and breathe. It would seem likely too, that they can hear. One cannot help wondering about the possibility of their smelling, tasting and touching – all of which are very somatic. It is also very likely that they can communicate very fluently telepathically. I had a dream recently in which I was worried that my mother was dying having forgotten that she was already departed, but I got a verbal message from her telling me in no uncertain terms to stop being foolish! Considering all this it would appear that belief is just as important, at the right time, as scepticism.

❖ 5. The universal conscious.

It has become apparent since I wrote the above account of my new model of the afterlife, that since mediums, truthful ones, can get access to the afterlife by using telepathy, there must, as written, be a conscious PP connection in addition to the unconscious one provided by the BPE, that is the UU, and again, as written, it must be symmetrical with the UU. It is essentially the collective conscious of the present life amalgamated with the CC of the afterlife and I call it The Universal Conscious (UC). If one thinks about it for a few seconds, it becomes obvious that, since the AL and the PL share the same space, unless there is something else operating, they certainly *do* have access to the same, well-known, familiar collective conscious, called for this occasion, the Universal Conscious, which is, of course, composed of psychoplasma, and which fades, at lower altitude, into the denser UU.

❖ 6. Precognition

Precognition is simply foreseeing future events. When one changes locality new sensory data is perceived, just as in temporal relocation things do change, although one cannot normally alter one's temporal speed and direction, as one can one's spatial speed, although subjective time does vary with mood.

It seems though, that in the AL the departed, unencumbered by the deadweight of the body, can travel, possibly instantaneously, from one spatial or temporal location to another; this has been established beyond doubt, but it may not apply to the future since the future is unreal. But we are talking here not of time travel in the AL but of precognition in the PL.(See diag. 13).

Diagram 13. is of world lines for the indicated objects, and shows simply how given entities relate to the dimensions of space and time. The space above the horizonal line is the future and that below it is the past.

C indicates the point at which two ships on a collision course will collide, assuming that their pilots have unknowingly miscalculated. This will constitute a potential event of which the GW will be aware.

All items of significant information in the unreal future will be potential events, seen only by the transcendental GW which is 'everywhere and everywhen', and in the event of their being relayed - in a dream perhaps - to a midshipman, will be registered as a premonition which may or may not be acted on by the captain, who may not know of the presence of the other ship.

In this diagram the future has not happened and is therefore unreal while the past has happened and is consequently real, leaving its evidence behind it i.e. ancient Roman and Egyptian statues etc. Precognition is deemed to have occurred in the case of the man who dreams the name of the winner in a horse race. This is shown at (a) or (b) where one of them wins. This too, is the work of the GW operating while the subject is asleep. The GW with its connections to other GsW in neighbouring unconscious minds, minds which never sleep, and its capacity to roam freely beyond time and space will know or feel, that the future is not yet real, but in its capacity as a knowing intelligence it will indicate points in space and time where events strong and significant enough to the subject will occur, i.e. precognitions, premonitions and similar events.

So we have come round to the GW again and its status as the mainspring of the paranormal. All is change except for the speed of light (physics) and the gods within - and without if there are any- (metaphysics). It is interesting that ever since the dawn of life there has been a psychoplasmatic, or at least bioplasmatic connection between every living thing and its neighbours which means that since then there has always been a tenuous mass of sensitive psychoplasma enveloping all of space and all of time (local to Earth) through which, presumably, the possibility has always existed of executing the (a) spatial and/or (b) temporal teleportation, that is, flipping to another point on the planet or another point in time, or both. It is, however, only now, by employing the scientific method, that these phenomena can be analysed and used, ultimately, to develop a technology to improve the lot of the human race and possibly all other life on the planet. I am, actually, fairly sure that many of these techniques are, and have been throughout history, employed by 'adepts' in the practice of the occult, but it is only now, with the advent of the astrological Age of Aquarius, that the unifying power of science shows the true scale of future possibilities.

So it is clear that the past exists and is concrete, insofar as it really has happened, and can be explored i.e.by archaeologists, living or dead, and the future is unreal which disqualifies the 'block theory of time, I think. If one could get 'outside time' (in spirit) as I know is possible from the result of my quasi- thought experiment with the afterlife, in which I became psychoplasmatically attached to Beethoven (seven lines further on and see chap.8 item 6 no.58,59,60.) one could get an objective overview of ongoing events in the past, present and future (except that the latter would not exist, I think) in the way that I got an objective view of the Sumerians and their cuneiform script and the raised letters on the iron plate of the water board, except that this view of the Sumerians, I believe, was only a faint shadow of what it would have been had I been fully existent in the AL. It was, essentially, I believe, as is all such motion, a case of the spirit, I myself in other words, directing the GW to take me near the object - the Sumerians - but finding it difficult to keep control – not being very practised at the activity, and aware at all times of the danger of becoming psychoplasmatically involved with the Sumerians. This danger could result in one of three unwanted situations, (a) one could accidentally resurrect someone, or (b) lose one's life by being dragged by powerful spirits into the AL, or (c)one could get into a messy PP to flesh interaction with some departed being, as I did with Beethoven, a kind of spiritual carcrash.

The actual action or mechanism of travelling outside time I think, between two points, could vary between instantaneous and continuous much as a pilot in his Spitfire could swoop and dive by using his 'joystick' and in this scenario travel would not be through BP, but would depend on the mysterious power of the ubiquitous god within, moving in a non-local universe as would phenomena described in chap.5 item.13. (ref.16). Travel would also not be limited to the years since one's birth but would in theory be limited only to such time as one would start to feel anxious and getting too far from home in 2015 AD. As already suggested one could surmount these obstacles by forming a band of scientists in the afterlife, united by a mass of PP to make a safer and more comfortable expedition, perhaps to observe ancient Egypt or Neanderthal Man and earlier. This, speaking from experience I know to be fact, but it could be done only from a base in the Afterlife, so one would need to be among one's departed friends and colleagues. The knowledge gained could be relayed to the present life by mediums using telepathy or, better still, resurrections. Until today I believed that symmetrically with teleporting, i.e. resurrecting, people from the AL, it should be possible to teleport into it but in the context of my new conception of the afterlife this is not possible because one would have no body there. It would appear that the only way of getting access to it is by the time-honoured mechanism of dying. Indeed if one could teleport into it one would still be in the present since the AL is coeval with the current one, I think, except for those spirits who choose to remain near home, the time and place of their death in the past. (See diag.3). It is, clearly, simply a matter of the physical body becoming defunct, so that the mind and spirit can float free.

At this point a 'thought-experiment' is indicated. If I, sitting here at my PC, were suddenly to keel over and die, I would, presumably, find myself in the afterlife, minus my body, and possibly with a number of friendly or related spirits hovering near, ready to greet me. The only difference, I think, would be that I had no body and could see, hear, feel or in some other way sense, the presence of the friendly spirits, of which I had no cognizance when I was alive. The strange thing is that things sensible when alive would remain so when departed, I think. I would still see the bookshelves in front of me, my computer and the nearby stack of CDs, just as in an OBE. I would also see my body lying on the floor. One wonders whether the other four senses would remain. Presumably I would then drift or fly away with the friendly spirits to some other place; I think the spirit remains permanently attached to its GW.

Thinking further on these things, it would appear likely that the spirit itself is not time-sensitive in the human animal; it doesn't change in essence or identity, only in size (with the growth of the conscious mind, the acquisition of knowledge and the involvement of the head), so it would preserve its identity irrespective of the time or place at which it existed.

But there is no such thing as time; no truly measurable substance as in the case of space, there is only the illusion of time through the counting of regular periods as in the vibrations of the caesium atom, which is used as the definition of the second. An animal has no concepton of the passage of time, I would think, although I am sure that they can be bored, excited, angry and even in love. For the human animal there is only locality and his/her state of being.(bored, anxious, enthusiastic, disappointed, busy, absent-minded etc.) The expression is not 'here and now' but 'here and what (state)' or here and where (in one's imagination) especially in a post-relativity universe. 'Here' is the state - presence - of the god within (the GW is never absent, whereas the spirit sometimes is, though not completely, as in the Near Death Experience, or 'distant in his web of thought' to quote a poet, whose name I forget, and 'now' only comes in when the head 'wakes up' and the conscious mind comes into play. 'Time' is an intellectual construct, in essence simply the sum of a number of periodically recurring events. So there is only 'place and state'. I am in my place and you are in yours. Your universe is marginally different from mine; you are in a different place in spacetime and the view therefrom will be different from mine. Our private universes overlap with those of other people to form the common universe as in a Venn diagram (diag.1 chap.1).

In chapter 4, under OBEs and consciousness, it is explained how the subject is able, while out of the body, to direct herself to travel into nearby rooms, through the roof to different parts of the environment and different parts of the city that she hadn't visited and could verifiably report what she'd seen there. This was because the spirit, the person herself that is, with the aid of the ubiquitous GW, is able to steer herself to any destination in its compass (ref.1a and 1b). I think, unlike the GW in psychometry and dowsing (following item) which knows, via PP I think, or feels, the spatial location of sought-after objects and reports them via the unconscious to the conscious mind, it can not know the nature and temporal location of events in the future, because there aren't any, though there are 'abstract' potential ones. This is where I disagree with the many scientists who believe in the block theory. I do not believe in the 'block theory of time' since the future is unreal and I see it as one more attempt to impose logic (symmetry) at the expense of feeling, thereby attempting to force Nature into a geometrically elegant but untrue framework.(See Primacy of Feeling). So: We may conclude, there are no real events in the future, merely a confusion of potential ones. All this neglecting the fact of recurrent cyclical events such as sunrise and the seasons, which themselves cause further cyclical phenomena, hence astrology. Therefore – no extended travel into the future.

It would be extremely interesting if a person who often has OBEs could be encouraged to try travelling into the past; some thing which is shown a little later, to be possible from a position in the AL. Unfortunately the person in an OBE is most definitely in the present life so this would be impossible. The assumption that the ubiquitous GW plays a major part in precognition seems practically certain and is strengthened by the fact of precognitive dreams. I did, actually, as written, try a quasi-thought experiment with the aferlife, which was simply to imagine being there, which with my afterlife contacts was fairly straighforward, although I wasn't observing very closely, and the GW stayed firmly in my gut, and try going back to the time of the Romans, then even further past the pyramids and thence to the geological periods, at which point I began to get a chilly feeling of getting lost, losing contact with friends at home and had to hurriedly retrace my steps. It was a very rewarding expt. however, and provided more evidence for my theory. I imagine that a similar thing might not happen if I tried a trip into the future, since the future is unreal and there would be nothing to see, except for a chaos of possibilities, or the GW would not oblige.

Here we have it then; the GW, as in psychometry, with its capacity to connect with and employ the gods within in its fellow humans – all six billions of them, to transcend time and space, simply feels and knows what is going to be encountered in the future of the organism. This would have good survival value for the organism for whose well being the GW, guided by the spirit, is always working. This, I think, is the obvious solution to the problem, despite its almost embarrassing simplicity.There is no getting away from the fact that the GW operates by what to us can only seem like magic. As a divinity it is beyond our comprehension and always will be. (Ref.1a).

We should not let this fact depress or discourage us, however, for what can be known about the workings of psi, which I think TMM enormously expands, simply opens up intriguing new vistas of discovery, analysis and exegesis, as well as the practical business of evolving a useful psi technology, which I see as being as revolutionary as that deriving from physics, and the forerunners of which can already be seen in the activities of Uri Geller (who is most certainly not a fraud, as seen and felt by me in a telepathic connection) and psychic mediums generally. The theory of material mind is simply the bare bones, the general framework, for decades or centuries of exploration, leading gradually to an integration with physics and, almost, the sought-after GUT which can, actually, only be a 'QUASI-GUT' since the activities of the gods will be forever hidden, I think, as just now stated. This will involve both physics and metaphysics, of course. But will that really be the end of it? Will that constitute a final state, perhaps of co-operative activities between peoples and nations? Who can say? I would hazard a tentative 'yes', as far as the solar system is concerned, but there are other solar systems, other planets, and I suspect that what is happening here is happening, has happened, and will happen, in thousands of other places in our galaxy and millions in the universe at large: and we are only in the infancy of the technology in our non-local universe, including journeys to the stars, I think, which will stem from TMM and the god within.

❖ 7. Psychometry or token-object reading.

Psychometry is the ability some people have of being able to derive information about a specified person from handling something belonging to that person, an article of clothing, a piece of jewellery, a book, etc.. It must be remembered that every material thing, although not necessarily involved in its generation, is permeated by the spiritual component of PP and the CC, hence an article of clothing must be connected in some way to its owner wherever he/she may be, if they are still alive, and even if they're dead, via the universal unconscious or the universal conscious.

At this point I would like to adduce the case of Eileen Garret, who was written about by Colin Wilson in hisbook, 'Beyond the Occult' from which I broadly quote (ref. 9a). Mrs. Garret was a famous medium much sought after by scientific investigators of psi, which included a psychologist, Laurence LeShan, who worked with her in 1964. He clipped a lock of hair from the head of his twelve-year old daughter Wendy, persuaded his neighbour to give him a tuft of hair from his dog's tail, and plucked a fresh rosebud from the garden. He then put each one in a clear plastic box. Then he retreated behind a screen with the boxes and Mrs. Garret had to put her arm in through a narrow hole. LeShan took a box and placed it where she could touch it. She immediately identified it correctly as the box containing his daughter's hair, and then went on to make incredibly accurate comments about the child. Her first remark was, "I think I'll call her Hilary – she'll like that". In fact when Wendy LeShan was four years old she had developed a crush on a girl called Hilary and begged her parents to change her name to Hilary. But the incident was long forgotten – it had not even been mentioned in the family for years.

Mrs Garrett then went on to make a series of weirdly accurate comments on Wendy – for example, that she loved horses and had recently developed an unexpected interest in American history. Her insights into the dog were equally impressive.........the neighbours had only recently moved in. Mrs Garrett said that it had a severe pain in its paw, and that it seemed to have a Sealyham companion......the animal had in fact cut its paw which turned septic....and dog fanciers said its bone structure reminded them of a sealyham........and the rose.....Mrs Garrett said the soil was too acid for it to grow well, something that LeShan had been told by expert gardners.

I think I can explain exactly how all this happened by referring to the TMM (fig. 4). First of all, it is clear that the conscious minds of Mrs Garrett and LeShan would certainly interact telepathically at such close quarters so that telepathy (or remote viewing) would inform her which of the boxes she was touching. As to the long forgotten information which Mrs Garrett divulged, first one must remember that the god in her and the god in LeShan, located at the solar plexus, would be reaching out to join each other via BP and that there would be a high degree of unconscious telepathy between them at such close quarters. The reader may remember that in the section in chapter three dealing with memory, it was revealed that the mid-term memories are stored in the heart while the oldest ones are stored in the abdomen which means that Mrs Garrett could thus have access, through a heart-to –heart connection to Leshan's mid-term memories, or through her unconscious mind to the unconscious mind of LeShan, and the old memories stored therein: Wendy's remarks about her former childhood friend is an example. Her friend would have been long forgotten but the memories of her would still be existent, stored as traces or neural circuits, (ref. 3) in LeShan's heart or abdomen, and Mrs Garrett would automatically give utterance to them as they came into her own conscious mind. The rest of the information disclosed by Mrs Garrett could have been through a combination of normal (conveyed by PP) and unconscious (conveyed by BP) telepathy and the activities of the gods within.

CHAPTER 7

Physiological functions and group minds

1. The function of the sneeze; 2. The function of the cough; 3. Clearing the throat; 4. The significance of the Burp; 5. Predictions made by the theory confirmed as true; 6. Discoveries and elucidations made possible by the theory; 7. PP bonding between people, animals and vegetation; 8. Group minds; 9. Blobs, funnels, tubes and streamers.

❖ 1. The function of the sneeze.
The sneeze, besides its physical function of expelling mucus or other irritating matter from the nose, also has a parapsychological function, namely, to expel PP from the nose and sinus cavities. The PP builds up as a natural mechanism and accompaniment to a growing expectation of some social greeting or acknowledgement as illustrated by the following account of an experience of my own.

 I was standing near the path in my local park when I saw a woman appear from a thicket about fifty metresaway. She saw me and began to walk towards me and I watched her approach gradually nearer. When she was within three metres of me I thought to do the natural thing and say "good morning" or some similar greeting but for some reason I didn't, at which moment she sneezed. The PP built up in the nose and sinuses was now superfluous, an embarrassment, so it was expelled by the sneeze thus clearing the head, and she walked on without speaking.

 An example where I myself did the sneezing goes as follows: I was shopping in the supermarket yesterday and my path crossed that of a pretty girl who looked at me in a friendly, inviting way. I did not say hello and she wandered off towards the end of the store. I decided to follow her and strike up a conversation, but a couple of minutes later, seeing that she was now too far away, I changed my mind and turned left instead of right. Immediately my nose 'tickled' and I sneezed a neat little sneeze. Once again the socially-oriented psychoplasma build-up in the nose was cancelled and ejected in a sneeze. This goes to confirm an observation made elsewhere in the essay, that during a conversation there is a continuous exchangeof PP between participants, which implies that a certain degree of close-quarters telepathy is involved. This is true of participants who are face to face, but in long-distance telepathic contacts, there is not an actual exchange of PP, but simply a slight flow in the direction of the person with whom one is communicating, just as in a flow of electrical current where electron (a) leaves the negative pole when the switch is thrown, but never reaches the positive pole because the the switch is made 'open'again, before it can do so. Despite this there is a brief flow of current, the lamp will glow, though the electrons issuing from the negative pole never reach the positive one. A better example is the corridor full of a flowing queue of people, all moving in the same direction, where the people at the entrance push the people in front of them who push those in front of them, who push etc. so that the people at the exit are pushed by the people immediately behind. It can be seen that the effect is thus transmitted over some distance even though any given person will not make it all the way from the entrance to the exit if not given enough time.

Another example drawn from my experience follows. I live in a care home and most mornings there are two or three lady cleaners cleaning my room, the corridor outside it and the rooms opposite mine. There is often a lot of light-hearted badinage between them, the other residents, and myself, and sometimes this can lead to petty rivalries between them, myself and the others. I have often been in my room aware of just one of them, in a kind of telepathic rapport. This may last for a few minutes and then, having forgotten her, I will start in on a new task or use my mobile to call someone. Instantly a loud sneeze will echo in the corridor outside. Clearly, she had not forgotten me!

This kind of thing happens to me several times a day and I always know that the sneezes are the result of unconsummated short-range telepathic communications. A failure to satisfy expectations. As already stated I think the mechanism must be a gradual build-up of PP in the nose and, possibly, sinus cavities which, if it were allowed to reach its potential, by the person causing it, would lead to a free-flowing conversation or similar positive social interaction, or, literally, 'meeting of minds'. I mention it because it is more evidence for the existence and behaviour of PP.

On re-reading this material a few months later, it occurs to me that in conversation there is almost certainly a distinct interpersonal flow of PP between the participants which would account for the telepathic communication between the conversing couple, (which would be about 10% I would estimate but variable) which is extra to communication via words and body language, as mentioned elsewhere in the essay.

❖ 2. The function of the cough

The function of the cough is similar to that of the sneeze but it operates at a deeper level in the heart. It is usually a consequence of disappointment or disillusion in person A who had been led to believe by person B that B would like to become more friendly, but who suddenly changes his/ her mind. There is still a solid build-up of PP but it is in the heart and lungs rather than the head, the heart being the locus of emotion, friendship, and love. The cough is associated more with solid friendships, their forming and dissolving, than the sneeze which operates at a higher level, and indicates light-hearted flirtatious, playful or tentative behaviour. Sometimes the build-up of PP in the heart and lungs can be the result of an accidental or uncomfortable friendship or association, and in this case the participants, let's say two people of either or both sexes, will cough repeatedly and clear the throat, to eject the other person's, as yet unwelcome PP. This situation, I think, would settle in time into a more comfortable relationship as its participants got to know each other. It can also be present at, say, a job interview where the applicant is having to relate to several unknown people. This actually happened to me in my interview for a place as a student at Goldsmith's College, University of London. For what it is worth however, I did get the place!

The cough too, of course, has a physical function, namely, the expulsion of irritating material from the throat, lungs and stomach.

❖ 3. Clearing the throat.

I think this is essentially the expulsion of PP which has tightened in and around the throat and windpipe, and which represents the entrapment of the spirit of person A by that of person B. This too, of course, may simply be the expulsion of excess phlegm which does not deny my argument since the physical and the metaphysical necessarily co-exist at all points in the body.

❖ 4. The significance of the burp

This displays the disappointment or phlegmatic acceptance by person A of a change of mind in person B, who was making gentle overtures of friendship to A. The burp originates in the gut, stomach to be precise, and comes directly from the GW. It happens between members of both the same or opposite sexes.

❖ 5. Predictions and observations made by the theory confirmed as true

- that a houseplant, and, one presumes, larger vegetation, will grow bigger, stronger and healthier, if it is kept near one or more other plants. This was confirmed by a statement made in his book on houseplants, by its author. He stated clearly that plants grown in close proximity will grow bigger and better thus confirming the prediction of the theory.
- That there are memories located in and around the heart. This was confirmed by a TV program broadcast on 26/06/05, on channel 4, which revealed that according to recent research, neural feedback circuits associated with memory had been found in the heart. This also supports my contention that the heart (spirit) is the foundation of the conscious mind.
- That there are also memories in the gut (intestines). This claim is supported by a section from the book by Steven Rose, the neuroscientist, 'The 21st Century Brain'(Ref. 14) in which he explains that there are almost as many neurons in the gut, as in the brain itself.
- That it is impossible to feel without knowing that one is doing so. If one has a multitude of feelings one will be strongly aware

of some of them, and perhaps only vaguely of peripheral ones. If, however, one has only one single feeling one will be aware of it, no matter how weak, and one will be aware of its nature and, probably, significance. Hence is, in part, derived the principle of 'to feel is to know' (TFITK) This does not need proof because it is self-evident.

❖ 6. Discoveries and elucidations made possible by the theory. (Discovered but, as yet not proven)

1. The existence and reality of the god within.
2. The Bioplasmatic Envelope.
3. The Existence and reality of the afterlife.
4. The duality of imagination.
5. The mechanism of poltergeist phenomena (Unconscious mind suppression phenomena UMSP)
6. The existence and reality of the spirit. (see addenda 115)
7. The Universal Unconscious. (equivalent of the BPE=Amalgamation of CU of PL and AL).
8. The Universal Conscious. (Equivalent to collective conscious of present life since this is shared by both PL and AL).
9. Teleportation and its mechanism.
10. Mechanism of PK
11. The individuality of the soul or spirit. There can exist only one of any particular spirit, but spirits can merge as is the case in true love. Psychoplasma can be male or female.
12. Revelation of reason for difficulty in replication of psi experiments.
13. The possibiliy of resurrection.
14. Probable solution to David Chalmer's 'the hard problem', the origin and mechanism of consciousness. This is the conscious spirit originating in the heart, which fills the chest, neck and head where it manifests as the conscious mind. The conscious mind originates and is based in the heart but at that location is much less focused and aware than in the head, mainly I would think, because that, (focus and awareness), among others, is the function of the brain. Also, I think, one tends to associate

consciousness with vision and hearing rather than the other three senses. These assertions are supported by the fact that the subject of an out of the body experience (OBE), remains conscious and usually hovers at ceiling height from where he/she looks down at her body which is lying on the operating table, or is unconscious for some other reason. The hovering entity is the spirit which, as stated, remains conscious and which, if the body dies, leaves for the afterlife. Since the hovering spirit remains conscious during all this it would seem reasonable to assume that the conscious component in the body, and therefore, the perceiving one, is the spirit (see ref. 2), which is normally contained in the body but which is excluded from the head when sleep occurs as the result of melatonin accumulating in the brain. It is odd that in an OBE the spirit sees directly, without the need for eyes, and that its vision is 360 degrees – all round - and this fact supports the idea that the perceiving entity in the waking state is the spirit (Ref. 2). Plato thought that the spirit was trapped in the body. Dreams according to this idea, would be the conscious spirit acting with and using material from the unconscious. The GW stays in and with the body during an OBE, I think. It is odd that dreams are experienced in the head, when the 'seat of consciousness' in the head has been disabled, and this I can't explain as yet.

There is a wealth of convincing evidence for the reality of the OBE, (see Chapter 4 item 10) and if one can accept this it is but a small step to accepting the reality of the spirit, which I know to exist and believe I can prove (see addenda 115). In an OBE something ascends, becomes aware of the surrounding reality, but gradually detaches from it; sounds grow softer, but the subject, for whatever reason, comes back to re-enter the body. If it did not the 'something' would become quite detached from earthly goings-on to go to the afterlife. Of this I am quite certain. There is, in fact, very little of TMM that I can actually prove as yet, although I can paint very convincing pictures which should make it easier for the reader to follow my reasoning, which is based largely on empirical phenomena (e.g. UMSP or poltergeist) and personal experience, as well as reports from the parapsychological profession itself, and hopefully, agree with it. I also intended constructing more replicable experiments before presenting the essay as complete, this, however, was before I had discovered the reason for the essential difficulty of psi experimentation (Chapter 2 item 2 and Chapter 8 item 10 and addenda 115

In support of my observation elsewhere in the essay, that the GW is always right and doesn't make mistakes in its dealings with its guiding spirit in the heart, and also that it probably doesn't need sleep (sleep on big decisions), but sorts out the material accumulated by the conscious mind during wakefulness, I would like to tell a little story about a recurring nightmare that terrified me for a year or two when I was about eight years old. The dream, essentially, was of myself rolling around and around at terrifying speed, inside an enormous globe. On the inside of the globe I could see a small circle of blue sky, and on its walls appeared a map of the world.

It is only quite recently that I realised the origin of this nightmare (possibly a guide to the origin of all nightmares?), and that was that I knew, at age eight, that the ground on which we walked was apparently flat but really curved since the world was in reality a globe. Unfortunately I thought we were living inside the globe and not on its surface, and found it quite impossible to work out how the sky managed to be so wide. Clearly, when I went to sleep at night, the GW and the unconscious mind found it extremely difficult reconciling these beliefs one of which was true and the other false, hence my night-time reeling and tottering around inside the globe as the GW tried to sort it out. The false belief was clearly in the conscious mind and it is clear that the GW in the personal unconscious finally sorted it out.

This, as it happens, suggests a very cogent explanation of why we have to sleep. (Assuming that this is still a mystery). My childhood experience suggests that the conscious, intellectual operations of the head are disabled by the accumulation of melatonin so that the conscious mind cannot interfere with the working of the GW in the unconscious mind, and its essential work of ordering and classifying material delivered from the conscious mind.

❖ 7. PP bonds between people, animals and vegetation

The phenomenon of inter-personal bonding is well-known, but it is not generally known that a material bond is formed which consists of psychoplasma. This is true of bonding with animals, which is also well-known and accepted, but it is not generally realised that bonding with plants is also quite normal and is of the same material kind. I think one can say that bonding is possible and occurs across the whole range of larger organisms on Earth, but only, of course, between friendly species. It is also a fact that since all life is connected, at least tenuously, by the BPE, at basal density, a very faint bond will already be existent, between any living beings, before any stronger, more specific material bond is formed. It may be argued, by opponents of the theory, that interpersonal bonds are simply memories, memories of one's friends, family, pets etc. but I would argue that I can remember, that is, there are loci of memories in my body, of every friend I ever had, but I am no longer bonded with them. I am bonded with current ones.

I had, until recently, a large potted money plant in my room with which I had formed a bond. I could feel the bond, and even see it very faintly because of dust in the air and of course, my plant was a friend. Plants do have feelings. One day a year or two ago, I was sitting in my room with the plant about ten feet away. I was wondering whether plants and other vegetal life possessed spirits: I decided not. On the instant I felt a huge 'whoosh' of what must have been transparent psychoplasma leap from the plant and onto my lap, and was aware of an unmistakable feeling of indignation from my plant.

On another occasion I had telephoned an old friend, whose number I had stumbled on in an advertisement having forgotten it until then. My call was not very welcome and soon she hung up. Just then a largish spider scampered out from under the fridge. I was still telepathically aware of the woman and her partner who, I could feel, were wondering whether to ring me back. Meanwhile the spider had stopped about a yard from my feet. I decided to try to teleport it to another part of the floor, and then to the back of my hand. Nothing. Dead still. Ten minutes later I sat down on a chair still keeping my mind on the spider which hadn't moved. A few minutes passed and then suddenly a girl at the local office of social security, whom I'd invited out for a drink, floated into my head.

On the instant the words, "lost it" came into my mind, meaning the telepathic contact with the girl and her friend, and at exactly the same instant the spider ran across the floor and under the wardrobe. These three events were obviously, I think, closely connected. The spider could feel a warm, benevolent presence, myself, the girl and her friend and when the telepathic contact with them broke it felt the change and disappeared.

A related incident happened to me some months ago on my way to do the shopping. It was a cold, wet, miserable day and as I picked my way past a garden with a low wrought-iron fence, I spotted an earthworm lying on the pavement. I stopped to look at it intending to put it back in the garden. I looked at it for about thirty seconds and suddenly, to my astonishment, I realised that I was bonding with it. I could feel it! I quickly picked it up, put it in the garden, and went on to the shops.

❖ 8. Group minds

Any group of people whose members regularly and frequently meet and intermingle will form a group mind. The smallest one usually encountered is the family mind, father, mother, sister, brother etc. although technically two people would be the smallest.

Any group mind will share a limited collective unconscious, that is, in the case of a family, each member, in addition to having a personal unconscious will share an unconscious that is the product of the tendency of the god within to expand, move outwards and connect with other gods within, thus producing a network of communicating unconscious minds, whose BP density will be greater than that of the surrounding unconscious mind(s). A similar constellation of unconscious minds will exist in a small community, a village or a town, and its conscious counterpart, the group collective conscious will also be present, taking the form of a large blob or system of linked masses of PP. (diag. 5.) One would suppose that in a close, say, family relationship, since the gods within reach out to join each other via unconscious telepathy, it would be impossible to keep a secret, but this would contradict the function of the god which is to obey and serve the conscious mind in all cases.

The concept of the family mind (conscious) was beautifully, and empirically, confirmed for me in autumn 2010 in my local park. I was sitting on a bench near the path, when I heard footsteps approaching from the right. I looked up and saw a man and his family coming down the path towards me. His dog ran up and licked my hand, while the man called out, "don't worry, he won't hurt you," and I replied "I know, I like dogs". I sat quietly as they passed, the man, his wife and children, a boy and two girls I think, and the dog. (Right now, as I write, I can feel the dog and the entourage reconnecting with me, as a result of my thinking of them, a weak, transitory bond).

As they passed I became aware of the fact that they were enveloped in a large, warm mass of PP, all of them including the dog. It was clear and of a light, honey-like consistency, but much lighter, only about five or six times more dense than the surrounding air. I was very strongly aware of it, and of its objective reality. I felt an urge to get up and join them but didn't do so. I looked up and I caught the mournful, lugubrious gaze of the dog looking back over its shoulder. Instantly, as its eyes connected with mine, I became enveloped by the family blob of PP and did not see but felt the man instantly become aware that somebody had joined his group. I felt his back straighten and his shoulders go back. I am afraid that instead of joining the little band I perversely remained glued to the bench. (Once again, as I wrote that, right here now, I felt the man's disapproval of my action in staying on the bench. My writing has brought the incident into his mind again, wherever he may be).

This, incidentally, suggests a direct though time-consuming method by which an interested party can procure empirical evidence of the truth of my theory. All such a party need do is to accept the truth of the concept of PP, the biological field and group minds and then patiently observe, hoping for an experience similar to the one which I have just now described. He should be warned, however, that it may take a long time. I myself first became truly aware of the existence of psychoplasma in 2005 but my first truly objective, that is visual, observation of a group mind came, as described, in 2010, five years later.

❖ 9. PP blobs, 'streamers,' 'tubes' and 'funnels'

As indicated above any collection, large or small, of individual living organisms e.g. a woman, two dogs, and a houseplant standing close together in the corner of a room, will be enveloped by a blob or mass of PP. The woman's husband may be busy in the kitchen and he will be connected to the blob in the corner, at all times, by a long 'streamer' or, to borrow from the literature of OBEs, a silken cord, of PP. It will not have the qualities of an actual cord but will be constituted solely of PP with vague, undefined edges as in the blob. It will also fluctuate in shape and density and change its cross-sectional area from about one square foot to one or two square metres depending on circumstances. In the shed at the bottom of the garden, the children, two teenage boys and a girl, observed by the household cat, will be busy tidying up. They too will be enveloped by a large blob of PP and from it a streamer of fluctuating cross-sectional area will stretch lightly house-wards to connect with the husband and the group in the corner. These will probably all be enclosed by a large dome of slightly more tenuous PP which encloses the whole house and its occupants. To extend this conception a town, to repeat something said in another part of the essay, but which is of particular interest here, will be one huge, formless mass of PP with hundreds of these other denser, smaller masses, that is, families, and blobs of various sizes, shapes, and streamers of various density and likewise will itself be enveloped by the less dense collective conscious. (Diag. 5.).

Elsewhere in this essay I have discussed the possibility of telepathic communication with the crew in the international space station, and have considered it somewhat unlikely. Recently however, while thinking about the possibility of a long diffuse, ragged 'streamer' connecting the ISS and Earth, it occurred to me that if the minimum density of PP possible, the basal density, is considered, the vague and fluctuating presence of a 'streamer' becomes logically impossible. Such things as streamers can operate only within the confines of the universal conscious, into which their diffuse boundaries can softly merge. As these thoughts occurred to me a new, very convincing image surfaced in my mind. It was an image of a polished, shiny, wobbly funnel or tube connecting the ISS to Earth, and I now see that this is how it must be, since, assuming that there *is* a connection, the density of the bounding surface of the tube must be no less than the basal density, estimated elsewhere in the essay as $D = 0.001$ of maximum possible density which is found in the body. This agrees perfectly with my first perception of the presence and nature of the BPE. The whole thing, funnel and BPE, can be visualised as having the qualities of a shimmering soap-bubble which extends the BPE boundary at that point into the stratosphere I think, and contains bioplasma and psychoplasma and, possibly, air. As stated elsewhere I estimate the general BPE boundary, excluding such things as funnels at about 8 kilometres above ground level.

CHAPTER 8

The Enlightenment and Beyond

1. Balance, the 'pendulum' and Rediscovery of 'The Old Knowledge'; 2. Witchcraft and the Occult; 3. Deism, Isaac Newton, and the Age of Reason; 4. What are we? Where do we come from? Where are we going?; 5. Evolution of life was, but no longer is, teleological; 6. Author's personal experiences of psi phenomena; 7. The quantification of metaphysics; 8. The vacuum wheel PP detector; 9. The umbrella PK demonstrator; 10. More on the non-replicable nature of psi experiments; 11. An example of mutually beneficial feedback between applied science and pure research.

❖ 1. Balance, the 'pendulum', and rediscovery of the 'old knowledge'; As Europe proceeded from the middle ages and its dependence on Aristotelian thought, through the Renaissance and its ideal of 'the universal man', to Galileo and Newton's 'Principia Mathematica' the triumphs of physical science and the scientific method came to dominate in society and the intelligentsia at large, and the 'old knowledge' astrology, alchemy, belief in spirits and hidden forces, sorcery, magic, the kabbalah and the occult in general faded into the shadows where it lingered until the advent of the astrological 'Age of Aquarius' circa mid – 20th century. This resurgence of belief, in the 20th century, is exemplified by the growing interest in alternative medicine and lifestyles, astrology, pagan religions and eastern mysticism.

Unfortunately, or fortunately, as the power of science grew, by virtue of its elevation of the faculty of reason, a strong reaction set in in the shape of the Romantic Movement, whose chief exponents were J.J. Rousseau and his dictum 'the heart has its reasons', and William Blake who coined the expressions 'the sleep of reason' and 'Newton's sleep', meaning that undue reliance on reason alienated humankind from its feelings, feelings which constituted the whole point of being alive for if feeling is absent there can be no life worth living.

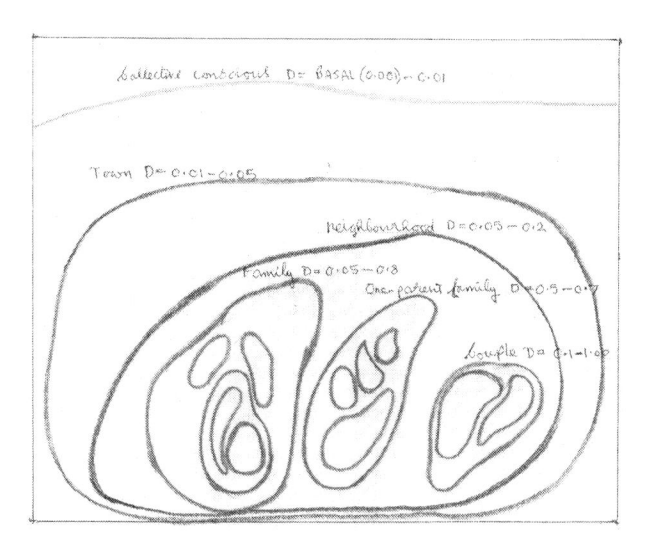

Diag. 5 Group minds. Contour defining regions of indicated PP density. All points on any given line will have same density.

My reason for these observations is that the foundations of culture tend to ricochet from one set of absolute beliefs to the extreme opposite under the pressure of social or scientific change; i.e. reason and science (logic, knowledge, power) versus romance and the arts (feeling, life, adventure) . It is clear, I think, that a balanced, stable society would ideally allow space for both of these world-views. Any progressive society would have its artists ***and*** its scientists in equal measure, giving credibility to both. This tendency has led to the current apotheosis of science and the total rejection by the average scientist, of the spirit or soul and all or any corresponding metaphysical possibilities. The failure to consider the reality of metaphysical entities can only doom to perdition all efforts to understand and organise the multifarious phenomena of psi. The swing of the pendulum must be slowed or halted and replaced with a balanced, harmonious system, in which the physical world and the metaphysical one are given equal importance, and shown to be complementary aspects of

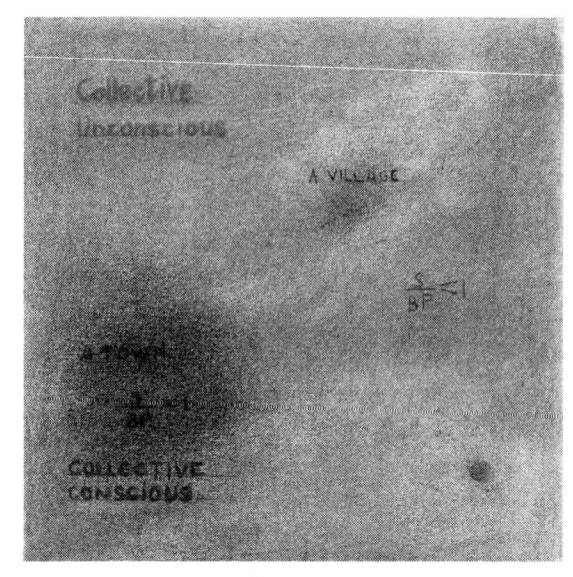

Fig. 10. Plan view of collective unconscious and
collective conscious at ground level.
Red indicates CU, blue indicates CC.

the whole. I do not, of course, approve of organised religion, which I see as an obstruction to progress, and a fundamental reason for the rejection, by science, of the spirit, consequent upon the house arrest of Galileo, among other abominations, such as torture by the inquisition for heresy or nonbelief, but see English witchcraft, or rather Wicca, its modern form, a pagan religion, as a thing of beauty having no bias against the body, an acceptance of sexuality, a deep respect for Nature, no creator god and no central law-making body. It goes without saying, of course, that the abominable practices of African witchcraft as brought to our shores by immigrants from that continent, and exemplified by the case of the child Victoria Climbie who was, in effect, ritually murdered by her guardians, can claim no moral or legal place here.

❖ 2. Witchcraft and the occult
Belief in the occult was widespread before the Age of Enlightenment, but was forced underground by the brilliant achievements of science. Many of its beliefs and practices were useful and soundly based in truth. I am not versed in the occult but am aware of some of its beliefs and

practices having had, in my youth, a couple of friends who were witches. The official name of the English variety of witchcraft is 'Wicca' whose adherents worship, or hold in high regard, instead of a creator, the 'Horned God' who presumably is Pan. One of the functions of the witch is to prescribe herbal cures for disease, infertility, impotence and any other kind of difficulty that besets the local populace. Many of the herbs are used today in the manufacture of medicinal drugs. He/she also casts spells with similar aims in mind. My own witch and his friend cast two separate spells for me both of which worked. One was aid to pay a debt and the other was to find a missing cat. The first required a strange ritual which involved the use of a saucer of water, salt, a piece of string stretched tight between my two friends' hands, and a small, many-thonged scourge with which the other witch whipped it, while the two of them uttered words from the language of Wiccan,the language of the English witches. The second was effected by the use of telepathy and a small copper talisman that I always carried with me.

My friend had made this talisman for me and had told me not to let anybody touch it which, presumably, would nullify its power. The cat belonged to friends in Paris with whom I was staying at the time. I wrote to him in England to ask him find the cat which he did. Apparently he had focused telepathically on the talisman and traced a path to an elevator where the cat was hiding. He told it to go home and voila! It returned bedraggled but OK.. These two successes could be coincidences, of course. Robert also told me that his father, who was also a witch I think, could stop a clock with the power of his mind. I know that this is possible having had similar experiences. (Chap.8 item 6 no.43) My witch friend used to talk, mischievously, of 'blasting' people he didn't like, but made it very clear that any evil act he committed would rebound to his own disadvantage.

The theories, beliefs and practices of the occult, furnish a rich storehouse of fascinating material for investigation, rationalisation, and incorporation in TMM. They are basically theories, practices and rituals (my witch friend preferred the word 'rite') based on esoteric knowledge of the world of spirits and unknown forces. The wide range of occult beliefs and practices includes astrology (for which I have discovered an 'a priori' proof which appears in the addenda), alchemy, divination, magic, witchcraft and sorcery. Devotees of the occult seek to explore spiritual mysteries through what they regard as higher powers of the mind.

The Western tradition of occultism has its roots in hellenistic magic and alchemy, (especially the hermeticwritings ascribed to Thoth the Egyptian god of the moon, and reckoning, learning and writing) and in the Jewish mysticism associated with the Kabbalah. The equivalent of Thoth in the Greek pantheon was Hermes Trismegistus.

Astrology was conceived in ancient Babylonia, prospered in the Italian Renaissance and oddly, (or significantly), in the 17th century Isaac Newton was still a believer, as was also Kepler a century earlier, but during the following two centuries, along with most intuitively acquired knowledge, it faded into desuetude, outshone by the brilliance of scientific achievement and the scientific method. Astrology, along with other divinatory and occult practices, gradually lost its attraction and came to be seen as superstition which was unfortunate, since it has its uses and can be shown to have a measure of veracity, as demonstrated by the proof mentioned a few lines ago. Despite all this destruction, however, the truth cannot be suppressed forever. In the Anthropic Cosmological Principle, Intelligent Design and The Theory of Material Mind, as well as the resurgence of Pagan beliefs, it can be seen that profound metaphysical truths are beginning to make their presence felt once again, and the searchlight of the scientific method is now being turned on the mysterious phenomena of psi. The Anthropic Principle, Intelligent Design and the TMM, represent the tentative restoration of the spiritual and metaphysical part of this 'old knowledge' which must, of course, be included with the physical knowledge acquired empirically, using reason to make a satisfactory picture of the totality of existing things and their interactions.

❖ 3. Isaac Newton, deism and The Age of Reason.

Isaac Newton, despite the mechanistic nature of the laws of motion and of his all-encompassing Theory of Universal Gravitation retained his belief in God and enjoined his followers to observe the design behind the mystery and beauty of Creation which, he believed, were such as to necessarily demand the existence of a creator. He claimed that although he had revealed a universe which looked like a great machine, it needed a creator to set it in motion, i.e. an uncaused cause. That is, his belief in God was founded on reason rather than revelation or the teachings of any particular religion, and his God, after he had started the 'clockwork' of the universe, no longer intervened in human affairs. This form of natural religion was called 'Deism' and it originated in England in the earlier part of the 17th century as a consequence of the gradual rise of science at that time, and was a rejection of orthodox Christianity. It gradually grew and by the late 18th century was Europe's dominant religion.

My purpose in writing of the Age of Reason is to demonstrate that it was essentially an interval separating former intuitive and spiritual beliefs (belief in the soul, the afterlife, witchcraft, 'scrying' [the use of a crystal ball]) from the present growth and re-emergence of New Age related beliefs (ref. 10b). The New Age is, of course, the astrological Age of Aquarius as written. This is the age of brotherhood and fraternity in which, it is claimed, priests will become replaced by scientists, and religion-based charities will be gradually replaced by secular organisations. In it humankind will become the steward of the planet, and achieve a kind of humanitarian objectivity or coming of age, in which responsibility for one's fellow-humans and a global vision will, (one would hope), lead ultimately to a democratic world-government. The operative words for the present but fading age of pisces are "I believe" whereas those for the age of aquarius are "I know" (Ref. 16). I see the difference of mindset, in the context of the theory of material mind, as the Piscean **belief** in 'supernatural' forces and entities, and the Aquarian knowledge of such things, now designated 'paranormal' or parapsychological; a knowledge which will eventually be rationalized and integrated with current scientific knowledge, and which, if I may be so bold, is already adumbrated in TMM.

❖ 4. What are we? Where do we come from? Where are we going? These words arise again and again, in the minds of the toiling millions, struggling blindly from day to day, looking for an answer. Some people say we are here to worship God, others that we are here to procreate and extend the race; I think most reasonable people would agree that in the absence of a clear and universally defined reason for being alive, one might just as well get on with it, the main point being to enj oy life with all its complexities, troubles and delights. One could say that the gods have given our spirits bodies to serve as media for the enjoyment of the wealth of activities, both mental and physical, that we are thereby able to indulge in (though Plato believed that the spirit was trapped in the body, an interesting idea when one thinks of toothache or smallpox). In order that life should be enjoyed, rather than endured, it must be rich in feeling and the good feeling, of course, must outweigh any bad feeling, while accepting that both will be present in any one life. Without feeling there is no meaning to life, and no such things as gratitude, peace, triumph over adversity, joy, ambition, love, humour, beauty, nobility, glory or happiness in the wealth of activities that humankind is privileged to enjoy.

The chimpanzee, despite its relatively limited powers of intellect and reason enjoys a natural, spontaneous and uncomplicated way of life since its activities are almost entirely based in feeling and instinct. Animals generally don't commit suicide, and even though many animals live in constant danger of being eaten by a larger one, they have few mental problems. They do, presumably, have rules for living, unwritten ones, although Nature is not always optimal in her working. I.e. the 'pecking order' observed by chickens, where each chicken is allowed to peck its subordinate, but must submit to pecking by its superior, but these rules have grown naturally through action and feeling rather than being consciously designed. They live from day today on feeling and instinct, unlike humankind whose lives are often stress-filled nightmares of plans, interviews, appointments, deadlines, traffic-jams, overdrafts, rules and regulations, calculations, data and bureaucracy, and, probably worst of all, alarm clocks! All logic-based activities. It is clear that the best course of action, where possible, is to live instinctively, and trust one's feelings.

Thinking people will never stop asking the questions, "what are we? Where do we come from? Where are we going?" and will never lose their sense of wonder at the mystery of existence, since, essentially, the universe is subject to the mysterious activities of the gods, which seem to operate both in living systems and inanimate matter (Ref: 9b), and the 'grand unifying theory' sought by physicists will be seen to be a will o' the wisp – science, and that includes parapsychology, will be forever incomplete. On a basic level this will be a parallel situation to that existing when Newton published his Theory of Universal Gravitation, since his system required the existence of a prime mover, a metaphysical being, to set the universe in motion, and the TMM also requires the existence of metaphysical entities, i.e. the god within and the spirit. Moreover it will eventually be revealed that the 'Age of Reason' aka the 'Enlightenment', was simply a period when just one of the faculties, reason, came to dominate, understandably though unjustly, at the expense of the others.

Now, in the age of Aquarius, it is time to set the record straight; mystery, magic, intuition and above all, ***feeling***, are revealed to lie at the bottom of the edifice called 'science', thus providing it with a firm foundation and the clear light of reason can take its rightful place as the ordering principle necessary to complete the whole.

❖ 5. Evolution was, but is no longer, teleological.

It is fairly obvious to most people, as concluded already, that life is to be enjoyed, and does not need any kind of transcendental philosophy to justify it or give it meaning. It is also common among thinking people to speculate about dying and the possibility of an afterlife, to wonder, as in the painting by Gauguin just now quoted, " what are we? where do we come from? Where are we going?" I myself am quite certain that there is an afterlife, as I have already described (chapter 4. Item 6) Meanwhile I have come to believe, since starting this essay, that there is, or rather ***was***, after all, a teleological aspect to the evolution of the species. I was imagining myself, as though looking down on myself from outside as in an OBE, living my life day by day, gathering stuff together to make life more comfortable and rewarding; food from the shops, articles for recreation, books, clothes, goods from other climes and other times, exotic fruits and antiques, and so on. And meanwhile, to paraphrase Omar Khayyam, "the moving finger wrote and having writ moved on" but on to where exactly does it move? Is the evolution of humanity random and open ended or is there a destination? While thinking about this I had a strong perception, a feeling, of incompleteness, a feeling that a destination was in sight, arrival at which would complete the process. It seemed reasonable to believe that this place was the Buddhist 'Nirvana' (extinction of individuality and absorption into the supreme spirit - not a creator God - but a spirit), where one would finally ' slip like a tear into the shining ocean'. I have not the faintest doubt of the existence of the spirit in the heart, and this fits in very convincingly with the idea of a supreme spirit, (not a god) a collection of the billions of spirits that must have lived on Earth to date. I should say here though, that since writing the above, (no matter how far-fetched and comical it may sound), I have been in telepathic contact with three ancient Greek philosophers who are still spiritually extant, in the afterlife, with their individualities intact (not absorbed) more than two millennia after their deaths, not to mention the seven million years old hominid which is in a similar state, which rather negates the idea of the supreme spirit.

Since writing the above it has occurred to me that the feeling described ten lines ago, which I felt to be an arrival at a destination, was exactly that but the destination was not a supreme spirit but my own final, clarifying discovery, the importance of which cannot be exaggerated, of the reason for the difficulties in replication of psi experimentation., namely, the mischievous god within. Having come so far in the last six millennia, 'natural philosophy', that is science, has now reached its final goal, the reunion of physics and metaphysics, (though that may be disputed by reactionaries) and in this respect it could be claimed that evolution *was* teleological but is no longer, being now open-ended and subject to the vagaries of discovery, invention and accident. This also implies abandoning the search for the 'grand unifying theory' much beloved by physicists, which is seen, now, to be impossible.

From now on, I think, scientific activity will be directed into vastly expanded areas of research and technology,materials, modes of transport, neuroscience and medicine for instance, but whatever their nature they will have to recognise that psi is no longer an earnest supplicant knocking at the door of established science, but must be accepted as a brand new, young and accredited science which is destined to produce technological wonders in combination with or transcending those introduced by physics and related disciplines. (see addenda 116)

❖ 6. Paranormal phenomena experienced by the author.
(1) 1955 Age 17 I was sitting in my 'den' experimenting with a 3-valve radio that I'd constructed, radio being my hobby at that time. As I cut

and soldered the words 'historical whodunnits?' came into my mind. I instantly guessed that it might be a case of telepathy, being already interested in the paranormal. I went into the living-room where my mother was reading 'Radio Times'. I asked her what she was reading about, she showed me the page. Halfway down the middle column were the words, 'Historical Whodunnits?', a programme that was having its first instalment that week.

(2) 1980 I was talking with a friend, a philosophy student, when he suddenly said, "That's odd! I'm thinking with my foot!" I knew that I was the cause of it being well aware that he and I were in some way joined, although I had not yet discovered the materiality of the mind and its implications. A continually recurring thought in those days was, 'all things are joined at the source.' At first I took this as a message from the unconscious but I later realised it could be coming from anywhere, since thoughts do occur at various places in the body and even outside it. It meant in any case, that one's mental experiences originated and were joined at birth. Interestingly, this suggests the astrological concept of 'time and place of birth', and the 'I Ching's' hexagrams.

(3) 1980 I had just then moved to a new address and was, for various reasons, very lonely mainly because I had driven away my friends as a result of the onset of a period of mental illness which took the form of an extremely idealistic and unrealistic attitude to society in general. I began to experience strange happenings almost immediately. One of them, very dramatic, concerned the timing mechanism on my central heating system. I would carefully set it before going to bed, to come on for two hours morning and evening. On waking up and getting dressed next morning I would notice that the flat felt cold. I would go to the timer in the next room only to find that the four timing spigots, which took considerable manual pressure to move, had all been moved back to zero! This went on every day for three weeks. I know that I don't sleepwalk, so that explanation is disqualified.

(4) 1980 Again, at about this time, I became plagued with a succession of exploding/popping light-bulbs in my kitchen. I was still living alone.

I would become aware of a 'black' train of thought, a light-bulb would begin to hum turning to a buzz and the bulb would suddenly 'pop'. This too went on for about three weeks.

(5) 1982-3 Still living alone in Finch House I felt everything was against me. I was losing my way. All relationships were bad, shaky. My landline telephone began to malfunction. I could not contact what I was beginning to think of as 'the outside world'. One day early in the morning before anyone else would have been there, I went down to my studio which was one of a complex of about fifteen others used by fellow artists. As I worked in my studio I felt a sudden need to go to the toilet. On returning to my studio I noticed at once that there was a small, slim wood-carving about a foot long lying in the middle of the floor, diagonally north-east to south-west. I put it back on the window sill whence it had fallen, making sure that it could not be dislodged by traffic vibration. Ten minutes later I had to leave the room again for some forgotten reason. When I got back it was there again in exactly the same spot!

I put it back and it happened again a few hours later, always lying in exactly the same spot. I felt that something was trying to tell me something, and now I see clearly that it was a clear case of 'UMSP'. My illness was forcing me into regions of unnatural self-imposed privation.

At the time, puzzled, I thought it was the god within, of whose reality I had become aware, telling me to be a bit more self-interested and to 'grab a little bit of all that's round you' to quote the Beatles, instead of being so unnaturally idealistic.

(6) */*/86 In my local pub a certain person (whom I, in my alienated state used to think of as 'the crab') could usually be found playing the 'one-arm bandit'. Even in those days I seemed to have unusual powers. I used often to go and stand behind this person to watch him play and every time I did this he began to win quite dramatically. There was also a girl there who often played pool. Whenever I turned my gaze on her she began to win. (She said that it felt as though a little angel was watching over her).I also had a friend who sold shirts in Greenwich Market. Whenever I joined him trade increased dramatically.

(8) 5/12/05 I became aware of the afterlife.

(9) 27/07/06 I dreamed and as I did so I saw into someone else's dream. It was a beautiful girl in her early twenties. We saw each other and then it faded. Then later something woke me up and I found a girl lying in bed with me, (the same one I assume.). I felt her arm and back pressing against me, real and solid. It gave me a terrible fright and I think it must have scared her too because she flew off through the wall near the headboard of my bed. I have, as I write this, realized who the girl could be. A very pretty girl who lives two hundred yards away from me, to whom I sometimes say hello. A similar thing happened to C.G. Jung, but, as luck would have it, his lady was old and deformed with only one eye which glared at him in a menacing and hostile manner (see ref. 1c). He too had the fright of his life and leapt out of bed to spend the rest of the night in an armchair. This experience was an example, the only one so far encountered by me up to that date, of a spontaneous teleportation.

(10) 10/01/06 Contact made with Rembrandt in afterlife. Non-verbal. I got the impression of an incredibly solid personality. Absolutely solid, real and substantial.

(11) 9/08/06 Made contact with younger Leonardo in afterlife. I have a picture of him as a young man dressed in flowing robes. I got a distinct impression of him as a charming and attractive personality.

(12) **/06/06 Took some pictures with my SLR but when I took the film to be developed it came out blank. I checked my camera and found it OK . This happened three times, three films wasted.. I've not tried since – too busy, but I suspect that it might be OK now since I'm under a lot less pressure and getting only a small amount of UMSP. (A later addition - 25 May 2011 - I recently used the same camera and the pictures came out OK).

(13) */12/06 Got into a frustrating 'PK based' relationship with my electric kettle. I was making a cup of tea,turned on the kettle and walked over to my table to finish some writing. Unfortunately I had got only

halfway across the room when the kettle turned itself off . I went back to turn it on again but as I approached it came on by itself. I began to walk away again but then it turned off. I spent the next minute moving to and fro' observing the kettle's antics, finding it impossible to move away without it turning off. In the end I accepted that if I was to get my tea I would have to stand near until it had boiled.

(14) */11/06 It was evening and I was boiling a kettle of water, tea in mind, when I found myself thinking to myself, "where's it at now then, mate?" Instantly there was a flash and a bang from my kettle, all the power points in the room became dead and the kettle was rendered useless. I am perfectly sure that my thought caused this example of spontaneous PK..

(15) */10/06 In the space of six or seven days the three lamps over my kitchen sink blew one by one, each one just as I came to the end of a particular train of thought and each time I knew it was going to blow.

(16) */ */06 Electronic bathroom scales malfunction. This has happened many times but has not happened for about a year now. There have also been cases with my pocket calculator, my CD player, and my mobile phone. This suggests that equipment was being affected at subatomic levels and interfering with electron flow which suggests, in passing, that an extremely useful and precise technology could grow out of TMM.

(17) 9/08/06 Another contact with Leonardo in the afterlife. Two actually, to be specific. The first was of him as a beautiful young man with charming manners (as portrayed in a picture of him that I have) in which I felt a mutual friendship. The odd thing was that later in the day I contacted him again but this time he was an older man of about sixty-five as in the famous red and white chalk self-portrait with flowing hair and beard. This makes one speculate about the nature of time, change and related matters in the afterlife. It must be possible for an organism to exist as the sum of its total existence from birth to death. Thinking further, however, this is how it must be at any stage of one's life, here in

the present one or in the afterlife. One simply grows to whatever stage of life and state of beingone has reached.

Yes. Of course! I see how it must be. Resurrection must necessarily always be of the subject as he/she was at the time of death, or of course, later. My 'special case' of the younger Leonardo must have been just that – a special case, bearing a similarity to my own sudden perception of my potential 'timeflip' to age 14, but essentiaally different since it was triggered by a picture rather than coming out of the blue as was my own timeflip.. (See 'retro-time-loop' diag. 4).

The difference here would be that in my timeflip, if I had done it, I would have been regressing to an earlier personal state of being, at an earlier time in the universe. So. I think it is clear that when I see the younger Leonardo by looking at the picture of him as a younger man, it is clear that I am seeing him at a time when the universe and every thing in it was younger than he as he was in the older red chalk self portrait, which meant that I could not perform a resurrection, since such can necessarily be performed only after the death of the subject.

(18) */*/04 I dreamed that a friend had once worked in Durham. When I saw him he told me that he used to live there.

(19) */*/03 For at least two months as I negotiated my way through the regulations and difficulties of hospital life, the words 'sub- zero' kept coming to mind, until one day when Stephen, a friend, turned up wearing a new sweater. Across its front appeared, in bold print, 'sub-zero'.

(20) */*/01 I was writing a poem in which I mentioned the name of a fellow patient who had been discharged eighteen months earlier. Two hours later I got a telephone call from him. He was living in Leeds but had remembered the name of the ward on which I was currently detained.

(21) */* /03 I was listening to the news, a lady announcer, and devoting I think, more attention to her than what she was saying. As I did so she suddenly lost track of what she was saying and came out with a whole load of gobbledegook. She stopped and laughingly said, "What

am I talking about?" I am quite certain that I was the cause of the occurrence. Such things have been happening in my life for a long while now, especially when I am watching TV. Sometimes I actually form *felt* relationships; I bond, especially with the newscaster delivering the morning news, making prolonged and steadying relationships, for them as well as myself. This is because I, with my gradually acquired powers, usually augment their feelings of well-being which, in its turn, makes me feel good. Sometimes I bring them down which brings me down, but if I relax a feeling of comfortable well-being supervenes. I only recently realised how and why this happens. It is because as explained elsewhere in the essay, whenever I think of or become aware of a particular person I automatically form a telepathic connection with them, therefore, since when watching the news I am automatically aware of the newscaster, I naturally connect with him/her.

This bonding also happens though not so strongly with recorded programs and films. I suspect that similar bondings occur whenever people watch TV, listen to the radio or even read a novel (between reader and author) although at only a tiny fraction of the strength of my own. The material mind is a far-flung web which surrounds and permeates all material things within its ambit

Precisely why I possess this power I do not know unless it is simply a consequence of my gradually acquired knowledge.

(22) 26/11/06 Tidying up, amusing myself by mentally addressing myself by different pseudonyms mostly nonsensical. I called myself "God" and a little voice said, "why not call yourself Philip, Philip? It is your name after all". So I called myself Philip while continuing to pick up bits of paper from my floor. Then I called myself God again after a short additional train of pseudonyms. On the instant one of the cluster of three lamps above mykitchen sink 'popped out' leaving only one for the other had blown the day before in similar circumstances. My thoughts on that occasion had been, "where's it at Cock?" Instantly the little voice (the god within I think – Bob Dylan's 'jokerman') had answered, "it's down to you, mate" or words to that effect, and the light blew, 'pop'.

(23) 1/12/06 I am becoming aware of a lot of new poltergeist (UMSP) phenomena, connected with my electric kettle which exploded a couple of weeks ago. It takes the form of a loud 'click' and it always happens at significant points in my train of thought. The phenomena seem to manifest, lately, more or less exclusively through electrical appliances, electric kettle, blowing light, mobile phone the screen of which split down the middle as soon as I tried to phone someone whom, for moral reasons I had earlier decided that I should not phone. Also, for about two years now, my TV has been making a loud cracking noise, as though its cabinet was being distorted under pressure whenever I go too far along a train of thought, or do or say something clumsy or inappropriate when in company. I have finally seen that my kettle clicks whenever I put myself into a mentally strenuous position, in a similar way to that of UMSP.

(24) 10/03/05 Transmission of a reel of sellotape through the wall of a cupboard in my kitchen. I quote here an extract from my journal:' I opened the centre cupboard to put them (bananas and oranges) in there, and to my astonishment there stood a jar of marmite with a reel of sellotape placed around its neck. How did the sellotape get there? I stood for half an hour considering all the possibilities. My conclusion was that it had somehow left its usual place in the top drawer of my chest of drawers, passed through the cupboard door, placed itself or was placed by hidden forces, around the neck of the jar. Inanimate matter does not move of its own accord so there must have been a psycho-plasmatic connection with some person or persons, possibly by Albert and/or myself, caused by the frenetic nature of the day's events. I took a photograph of it and went to bed that night, rejoicing. Amazed and delighted. My central theory was being corroborated and amplified in greater and greater detail.

It would appear appropriate at this point, however, to indicate the alternative to my final interpretation of these events. It is quite possible that the phenomena were produced in their entirety by 'normal' or 'common or garden' poltergeist activity. i.e. open top drawer, take hold of sellotape and carry it to cupboard, open door and place it round neck of jar, close door, but I had a strong, immediate intuition or conviction that this was a case of inanimatematter passing through other inanimate matter, so I assumed that this was what had happened. This does in any case happen, as the appearance of a girl in my bed and her subsequent disappearance through the wall clearly demonstrates. This intuition was corroborated by events which happened some years later, and which will be described later in the essay.

(25) 11/03/05 Another astonishing event. Having done my morning ablutions and had a cup of tea, I opened the door of my wall-cupboard space 'C' where my files were kept on the top shelf, I wanted to check some record, I've forgotten what, and reached up to get them. My eye was caught by something on the bottom shelf. A jar of Marmite. It was the jar of Marmite but now with no sellotape round its neck. Again I thought long and hard checking out the possibilities. My conclusion; it had passed from space B through the dividing wall, and into space C. This could still have been UMSP, but for the same reasons described a few lines ago, I concluded that it was a transmission of solid matter through other solid matter. One wonders did the jar disappear suddenly at 'A' to reappear instantly at 'B', or did it perform a slow, gradual transmission from 'A' to 'B'?

(26) */*/06 At many times during 2006, I have made, accidentally in the main, telepathic contact with various historical figures in the afterlife. I remember my first contact with Beethoven, my first hero in my youth. I felt a slight sensation of something tearing in the space above me. It was something fixed and immovable and I guessed at once that it must be Beethoven and in the next few days I learned how to make better contact with him. My first totally successful and satisfactory contact was of a warm, sunny benison and a vast and powerful personality.

Other figures contacted include T.S.Eliot whom I felt I was disturbing. He seemed sick and deep in depression, Michelangelo from whom I got a black, cool if not cold, but not unpleasant feeling and a feeling that something was being with-held. I think it was a hesitation, a reluctance to reveal himself completely.

I also made contact with Newton, Haydn, Bertrand Russell, Arthur Koestler, Einstein to whom I revealed, by accident, my ideas concerning time and made him, oddly, scratch his head. (I later discovered that he was already familiar with them through a friendship with Kurt Gödel the mathematician). I contacted C.G. Jung - an excellent feeling – and had a kind of feeling-based non-verbal discussion about some issue, I forget what. Picasso came over as a warm, soft, friendly, incredibly nice person, full of feeling but, on reading his biography a few days later, I learned that he had a very egotistical, cruel and selfish side to his nature. There was a colour associated with him, a soft brown colour which I have noticed in normal telepathic contacts indicates friendliness. I got a very good contact with David Hume the philosopher who, in his work, abjured and condemned all things spiritual or metaphysical as meaningless nonsense. He seemed slightly amused though not embarrassed to find himself in a place whose existence he had disparaged in conversation with his fellow philosophers. Coming now back to the present. Several times in the past year 2007-8, I've been in touch with Tony Blair and at other times with George Bush. I sometimes contact Tony Benn and on other occasions I get Prince Charles. I watched his wedding to Camilla on TV and was worried that I might spoil it. There was a strong contact but it went very well. I used to like watching important boxing matches, but the last time I did so I began to bond with the boxers themselves, and found that I was becoming part of the action so much that I began to hinder the efforts of the favourite, so I turned the TV off. I learned later that the favourite had lost!

The most astonishing contact of all happened while I was thinking about paleontology. I have a newspaper cutting which shows a reconstruction of a seven million years old skull of a hominid. The skull had been covered with a layer of clay or a similar medium and an imaginative model of the creature had been made.The amazing thing was that as I looked at it I suddenly realized to my astonishment that I was coming into telepathic contact with it! My first thought was that I was communicating with something living seven million years ago in the distant past, but I quickly realized that the lady, (I got the feeling that it was female), was in the afterlife right now, so she could have been anywhere in space, even, I had to admit, in some other 'dimension' although, generally speaking I don't like the idea of other dimensions, unless they are simply used as heuristic devices. I should say here, that my most recent and cogent conception of the afterlife, dispenses with the idea of parallel universes and higher dimensions, and places the afterlife exactly congruent, spatially, with the current one (see chap.6 item 5), and writing now in Dec. 2010, and having in mind my new conception of the afterlife, I suspect that although space in the AL is congruent, that is, **shared** with space in the PL, time in the AL is very different from time in the PL as suggested by the continuing existence of the Greek philosophers and the hominid from 7 million years ago.

What I have seen is that space, though not time, is basically the same as it is for the spirit in an OBE. That is, the spirit, when in the AL, remains attached to its own particular GW which, when directed by the spirit, will carry it to anywhere and/or, theoretically, 'anywhen', at any speed, I suspect. This is possible by virtue of the assumed or axiomatic fact that a divinity can transcend space and time – 'time' in any case, does not exist, only change. I share this belief with at least two eminent personages i.e. Albert Einstein and Kurt Gödel the mathematician. I would spend more time on this subject but the need to finish forbids.

One thing is certain and that is that my parents and departed relatives who have all passed into the afterlife, are definitely not defunct. They are living and active in some way, as was incontrovertibly demonstrated by the fact that they are definitely connected with and permeated by, the BPE. I felt this strongly with the first connection that I made, with my mother on 5/12/05.

(27) 2/05/06 I got into a heart-to-heart telepathic (telecognitive to be exact) connection with the girl living round the corner to whom I sometimes say hello. I learned that her name was Glenda but when I next met her she told me that Glenda was the name of her friend who was usually with her when we met.

(28) 31/07/07 Watching the news read by Moira Stewart and admiring her cool, formal beauty. She lost track of what she was saying, stopped to regain her composure and read on. I know that I was the cause of it.

(29) 10/07/07 Contact with Osama Bin Laden. Eyes only.

(30) 24/06/06 On the way home the words, 'eyes like saucers' came into my mind. I associate these words which I've had in my mind on other

occasions, with the girl who lives nearby and to whom I say hello. A few seconds later the girl herself and her friend, another girl, came up behind me and spoke to me. Some part of me must have become aware of their presence via the 'Biological field' (BF) which was described earlier.

(31) 8/07/06 More strange, longstanding telepathic experiences. As described earlier, when I watch TV I usually find myself getting close, attached to the people in the programme. This is particularly evident in the case of live programmes because I can see the reactions of the participants, but it also happens to a lesser extent with film and recordings. For some time I couldn't see why this happens, but I now think a possible reason is that the TV personalities and myself are, of course, connected by the CC which permeates through all obstacles between us, and, having seen clearly that the expression, 'the eyes are the windows of the soul', is literally true, when my eyes meet those of one of them, we are immediately in mental and emotional contact. I can see into their eyes but simultaneously can feel their reactions via the CC. Total contact. My contacts with TV personalities, especially regular ones such as those with newscasters, as explained, have become an important part of my life as I must have in theirs. I think that they must be aware that someone with special powers is watching them, not in a malicious or voyeuristic way, but as someone who enjoys the programme. Right now I find myself in touch with Adrian Chiles, a presenter on 'The One Show' on BBC 1, and I have a feeling that I am bothering him, so I think of someone or something else.

(32) 14/09/06 Contact with Leonardo in the afterlife. A small, dimly-lit room, two figures, Leonardo on the right in flowing grey robe. He and I drifted together and I gave him a hug. I felt the solidity of his body. (can this really be true!?). I think this may be a dream – not sure.

(33) */09/06 Contact with Jesus. I became aware of a radiant light, (sounds hackneyed but it's true) and I adopted a kind of reverent attitude (how embarrassing) but he soon made it clear that this attitude was inappropriate and that we were equals, which I saw at once was correct.

(34) 2/10/08 Contact with Hitler. The sublime to the….? Hardly ridiculous. I remember that before going into hospital many years ago with a mental illness, from which I have long since recovered, my thinking was very idealistic and praiseworthy but erroneous and wrong-headed. During this period I used to get a feeling (akin to 'delusions of grandeur') of proud self-esteem. It felt brown in colour in some way (brownshirts?) and was what I imagined a nazi would feel like. In my bathroom just now I suddenly felt myself in contact with Hitler and it felt the same. An identical feeling. Presumably he felt the same. Also, the GW being intrinsically good, would recognise the unnatural features of Hitler's beliefs, which were based on an unnatural ideal, just as it did my own, when I was ill, and they would manifest as the 'brown feeling'.

(35). */8/08 I was standing in the garden. Suddenly the name of a woman came into my head. Ten or twenty seconds later I heard her voice thirty feet away in the building. There was no visual contact.

(36) 6/11/08 Watching the news read by Riz Latif. I was in fairly close contact and admiring her silken blouse. As she finished and said goodbye, I looked at her face and she said, "that's all, bye Phil". She uttered my name! (As I typed these words iust now I suddenly felt her annoyance, and found myself in TP contact with her again.)

(37) */03/03 I was having an impassioned conversation with a girl in hospital. My thoughts were racing ahead and I thought in my usual direct and down-to-earth fashion, "f…..a……s" (two expletives). I had never used these words before, or even heard them so used, but a few seconds later the girl used them to express surprise. An example of close-quarters telepathy.

(38) 27/03/03 Another example of my thoughts 'escaping' into the environment. I was thinking about an aunt whom I used to visit as a child. She always greeted me with the words, "hello Cock" which was a common form of address in my home town at that time. As a result I began to address myself in my thoughts, here in my room as 'Old Cock'. I would think, " time to get moving, Old Cock". A day or two later I

heard a man on the radio address an interlocutor as "Old Cock". I had never before heard someone address anybody in this way.

(39) */*/06-7-8 I have had many contacts with departed pets of mine, (cats, dog), and felt at the time that it was quite within my power to resurrect them, teleport them to the here and now, but as usual, I was loath to do this since it might have harmed them, or vitiated the circumstances in which they normally live in the afterlife.

(40) 17/12/06 In the care home where I presently live there is one particular relationship with a member of staff which I find very difficult to deal with. Sometimes the feeling gets very bad and I lose my temper and end up feeling angry, frustrated and impotent. Tonight, however, somewhat untypically, I decided to let things be, 'keep it sweet', try to maintain a friendly feeling and be more patient. A couple of seconds later I looked down at the working surface near which I stood and was startled to see my stainless steel Chinese kitchen knife lying there. I had been searching for it 'high and low' the day before and had concluded that a nurse had taken it to the kitchen. I then realized that it was the work of the 'good ghost', (the GW) helping me, (because I had controlled my feelings, and been good myself) just like yesterday when it turned the light on for me, and the other times in the past when it had helped me or warned me.

This is a sort of 'inverse' UMSP, helping rather than playing tricks, and I think that it must have been the result of a collaboration between the GW of the 'difficult' member of staff and my own GW which acknowledged each other's goodness and got together to produce the knife. It should be said that the knowledge of any given GW may be limited to, and operates within, the confines of the person to whom it belongs; this person's knowledge and experience will be limited, but since any particular GW will be connected, via BP to the six billion gods in the guts of other people, it will have access to the knowledge and power of the entire population of the earth, and moreover, in its capacity of divinity, even more distant hidden regions beyond space and time.

(41) */*/08 I was sitting in my room thinking about the differences between human and plant-life. I had been wondering whether plant-life could know as well as feel, and had decided that the capacity of knowing required a spirit in addition to the god within. With this in mind I decided that plants didn't possess a spirit and instantly I got a great, indignant 'whoosh!' of BP and spirit (one presumes), or simply PP, from my plant near the window onto my lap and into my heart. I have discovered since that 'to feel is to know' means just that, since if one is feeling something one necessarily knows something i.e. that one is feeling. One can conclude from this, quite apart from experiments on plants which confirm the fact, that vegetal life does have a degree of cognitive ability (ref.6) and on the evidence of my plant's behaviour, do have a spirit, which, thinking further seems to be obvious since when they depart for the afterlife, the only thing which could demonstrate plant A's identity from that of its neighbours would be its spirit. I say this because, although I think the god within also goes to the afterlife, (it is a god and gods, I would expect, don't die), I believe also that all gods within are very similar, if not identical as far as distinguishing personality characteristics are concerned, unlike the spirit in the heart and its associated memories, which is the essence of the individual's personality.

I think, therefore, that plants do have spirits which must be distributed throughout their physical structure, in much the same way, perhaps, as the constituents of a hologram, tiny facsimiles of its main image, are distributed to make up the whole.

(42) 14/02/09 Just an hour ago I turned the radio on fairly loud to a music station, as a precursor to having a bath. I could hear it from the living room while I was in the bathroom. I stayed in the bath longer than usual, feeling in need of a rest from stress, and I became sleepy. Eventually I got out of the water, towelled myself dry and went into the living room. I went to turn up the radio and then suddenly realized that it should be fairly loud already. I 'twigged' immediately. A UMSP had occurred (poltergeist), and it had been turned down to almost zero! I am not sure why this happened but it could be related to my 'sleepy-head' feeling.

(43) 17/09/09 A milestone! My first success at making a physical object move by the direct power of the mind. There stands on the chest of drawers in my room, an 'umbrella PK demonstrator' as described in the main text (chap. 8 item 9). I got up this morning, sat myself down in an armchair in front of the TV set which stood on the chest of drawers. The PK demonstrator stood a few inches to its right. As I looked at it it began to rotate at about 0.5 revs./second anti-clockwise, and kept it up for about five minutes. Some of the time it was as if it was rotating itself. I just sat there blearily waking up, looking at it. I made no mental effort, I just sat there watching it happen. There were, of course, no draughts and it was out of range of my breathing. Also I had checked on prior occasions that the TV didn't affect it.

(44) 21/09/09 Today at 6:15 pm I was putting the finishing touches to a paper I had written, when I suddenly realized, just like Cleve Backster and his plants, that the letters were appearing before I touched the keys, and for ten or fifteen seconds I just hovered over them and watched the letters appear as they came into my mind. I did occasionally touch a key to keep it going but felt that I was floating above the keys, on a soft cushion of feeling, watching it happen.

I think that this happened because a few minutes earlier in the courtyard of the care home where I live, I had been to a party where they were playing loud Reggae music, one song of which was 'Oh! Baby, Baby it's a wide world' and the beautiful music had filled me full to overflowing with good feeling - in other words, strong and buoyant psycho-plasma.

(45) 4/3/09 For the last two weeks I have been experiencing a succession of astonishing coincidences, two or three every day. The final one came this afternoon. I was sorting through possessions when I found, to my delight, the small copper talisman described in chapter 8 item 2, given to me by my friend Robert the witch. I had not seen it for many years and had assumed that it was lost. A related coincidence had occurred a few hours earlier (or later, not sure which) while I was reading a newspaper. I suddenly came across a subheading containing in capitals the name 'Vulcan'. Vulcan was the spiritual name of Robert!

(46) */*/86? I was walking home feeling somewhat confused about shaky relationships, when a little voice in the air about 6 to 12 inches to the left of my left shoulder said "give a ghost a chance". I knew for sure that this was a telepathic connection with a true friend whom I had almost forgotten. He must have been feeling stressed and uncertain himself about my failure to contact him. The use of the word 'ghost', I think, merely meant that the PP attachment between us was on the point of fading, ghost-like, into nothingness.

(48) 27/04/10 I was standing in the dining room facing the TV. A fellow-resident walked in from the garden, passed behind me to go to his room closely followed by another. One or two minutes later, to my surprise, the first one came in from the garden again, passed behind me and went to his room but I know for absolutely certain that he didn't go behind me to get back to the garden so he must have gone from his room to the garden which was impossible because both relevant doors were locked. A minute or two later he reappeared to sit in front of me in an armchair.

He looked a bit disturbed and I think he said he was going to report me to the police (this might have been on another occasion), but either way he looked very strange, sat upright, arms and hands held before him like a dog 'begging', limp wrists – very much out of character. I think this could be the second spontaneous teleportation that I've witnessed, the first one being the girl Jasmin from round the corner whom I found in bed with me on waking suddenly one night.

(49) */7/10. I have become extremely sensitive to the biological field of vegetation, especially when the vegetation is dense. Walking in the park a couple of days ago, in bright sunny weather, I stopped on the stony path in the shade of some elder trees and large bushes and as I stood there I began to feel a strong, friendly bond growing between us .I felt that it enveloped me, and as I write right now, strangely enough, I can feel myself gradually coming into contact with them. Unfortunately something made me leave them, I forget what. It would have been nice to stay longer and find out more. This brings to mind the current 'new age' practice of hugging trees, which could, I think, be quite beneficial to the human system, since it would involve sharing PP with the tree, which

is an extremely strong and stable living system. This represents a belief arrived at via intuition i.e. it fits in with New Age thought – it feels right.

(50) 17/8/10 I have recently noticed that if, for some reason, someone is angry with me, I feel their anger rather than interpreting their body language.This, obviously, is due to a telepathic connection. I think this probably applies also to other feelings, pleased, surprised, amused etc I must check this with friends; do they feel the same?

(51) */8/10 I was in my room working at my theory when I suddenly noticed that I felt cold. I then realised that the reason I didn't feel cold before was because a friend was working in the corridor outside my door and her presence (and PP which we had been sharing) had kept me warm. When she went off duty I suddenly noticed the cold.

(52) */*/08 I was standing on the corner after a walk, in quiet contemplation. Suddenly the name of the pretty girl to whom I sometimes say hello, came into my mind. I turned round and she was standing behind me with her friend.

(53) */*/07 I was standing in the garden when, for no particular reason, the name of a certain woman came into my mind. Ten or twenty seconds later I heard her voice issuing from a room thirty feet away, inside the building. There was no visual contact.

(54) */*/08 Watching TV. A TV interviewer was talking to a retired professor of physics and his wife, about his new house and garden, I think. He was complaining about the noise from passing traffic which spoiled the peace of his beautiful garden, and instantly I had the idea of using a microphone, amplifier and large speaker system, which could cancel out the incoming sound waves by sending out phase-reversed amplified copies. At exactly the same moment my eye caught that of the professor and I know for certain that, as a result, he too saw the idea.

(55) 29/9/10 A mad, frenetic day. At one point I was surrounded by whirling PP and turning, saw, in the middle of my room, chatting and smiling, two young brown skinned lads, about 17 years old. There's not the faintest doubt that they were spontaneous teleportations from somewhere local, I think. It gave me a fright and I shook myself and managed to make them go. They disappeared. They didn't seem surprised but it only lasted for one or two seconds so they probably didn't notice.

(56) 30/9/10 Another mad difficult day, and another astonishing psi event: I spent some of the morning gleaning a little peace of mind playing a guitar that is used by residents here. A few minutes later I put the guitar down and went to my room upstairs to eat. Arriving there I stayed in my kitchen preparing my dinner. Five or six minutes later I turned to lean against a working surface while it cooked..

To my astonishment the guitar was standing, leaning against the bookshelves! I did not put it there and nobody else could have done so because I'd locked the door of my flat. It was obviously a spontaneous teleportation. An interesting aspect of the phenomenon was that the guitar must have passed through three closed doors and/or a few walls and ceilings to get there.

This occurrence clears up the doubt about the sellotape and the Marmite jar, mentioned earlier in this section, i.e. solid, inanimate matter can pass through other similar matter, in seconds. It does, however, raise another question, namely, how fast did the guitar travel? And another, was it a fast but gradual movement from A to B or did it disappear in an instant at A to reappear instantly, or rather, simultaneously at B? The latter possibility brings to mind an account of a certain Russian Professor Podshibyakin who reported, if I remember correctly, that he had observed in a human being, that certain changes, in the brain I believe, occurred in synchrony with sunspot activity, which suggested that the sunspot activity was directly affecting the working of the brain. This had me puzzled for a time since I know that my theory is practically 100% correct and I could not believe that the Earth's psychosphere could extend more than a few miles into space. It finally dawned on me that the solar activity could have been causing atmospheric disturbances, electrical storms and such which, in turn, could affect the psychoplasma of Earthly minds!

(57) ?/05/10. It was a sunny day in the garden of the home I used to live in. There was a person living there who, though mentally disturbed, that is, trapped in a private world, was essentially a very good, well-centred person who needed, instead of the medication fed to him which he hated, a trustworthy friend on whom he could rely. For various reasons I myself could not fulfil this function. Sometimes I would accidentally drift into a telepathic relationship with him and he would start uttering words, saying things which had very definite significance for me. One of them was "…. the lion…the lion's here…" which signified the astrological sign of leo which had great significance for me. Another time he began to utter the words of a poem I'd written many years earlier. They were "…… elmer nin sono…..do you know, no, no" and others which I've forgotten. This I took as an example of 'telecognition'mentioned earlier – direct transmission of words from heart to heart, or divination.

(58) 30/1/11 This experience I include partly because it made me laugh. I was typing at my PC when I experienced a short attack of flatulence which I put down to the baked beans I'd eaten earlier. It immediately occurred to me that this common effect of eating beans is what Pythagoras meant when he advised against eating them! I used to think, in common with others I think, that he connected the consumption of beans with some profound metaphysical truth. But now I saw that he had simply played a joke! I immediately found myself in touch with him in the afterlife and I heard, mentally of course, what can only be described as gales of Homeric laughter. They came of course, from Pythagoras who must have responded when I thought of him.

(59) */2/11. I was reading a book on philosophy and happened upon the well-known quotation from Heraclitus of the supposed fact that one cannot step twice into the same river. I instantly disagreed believing instead that a river is defined by its banks, not the water. I immediately felt a strong sensation of defeat and depression from Heraclitus with whom I had connected, but then, to my delight, and his also, I realised that the banks too would be changing, wearing away, so he was right after all! This brings to mind the observation, a product of 'the sixties' I think, that 'change is the only constant'.

(60) */2/11. On the same occasion, I think, I was reading about Plato and disagreeing with something he believed to be true. I instantly felt a sharp stabbing pain in the region of my heart and gut. It was Plato registering his anger. Of that I have not a shadow of doubt, and soon, although I am not yet in total agreement with him, I found myself agreeing more and more, and more and more admiring his perspicacity in describing and asserting the reality of his ideal forms. I got the impression that he was a much more aggressive and 'sorted' personality than Heraclitus who seemed more vulnerable. This amounts, in passing, to evidence that the AL can indeed influence the PL, i.e. that there can be intercourse, telepathically or otherwise, between the two.

(61) 21/9/11.(Yesterday) I heard on the radio or TV of a little girl in America, eight years old, who'd asked some very important person whether there really was a Santa Claus. Today 22/9/11 at 1:00 am, I looked up Chuck Berry in the 'quotes' section of my Seiko electronic dictionary. He wasn't there but in his place appeared Francis Pharcellus Church (1839-1906)(of whom I'd never heard)....and the following: "Yes, Virginia, there is a Santa Claus" replying to a letter from eight-year old Virginia O'Hanlon. Editorial in New York SUN 21/9/1897.

(62) 1/9/11 or thereabouts, I was in my kitchen and something reminded me of the kitchen of my flat in Coventry where I used to work fifty years ago. To my surprise it felt as if a I was actually there again, I could almost see the round table at which I dined and see the building worker who shared it with me. I began to feel that there was a danger of getting 'lost in the past' so I rapidly returned to the present. Just now, as I recalled these events, I remembered a doorway that stood behind the table and began to feel a desire to go through it. Quite soon I was getting alarming feelings of being in the past and about to enter the doorway, at least my head was – the rest of my body was firmly in the present. I'm afraid, once again I pulled back from the experience as I usually do. Perhaps when this essay is complete and out of the way I shall feel free to 'take a trip' to wherever I am lead. This experience suggested that such things were still in existence somewhere, but where? In in the physical material of my head? This experience requires more thought since it seems to suggest

that very old memories, *all experience,* may be stored in the head, and related ones in the whole body

(63) ?/12/11 I was distributing leaflets, on foot of course, for a vegetarian charity which I support. As I came nearer to, within about 30 or 40 metres, of a beautiful victorian house of the sort that gets divided into flats, I suddenly felt the presence of a certain person whom I had last seen in my previous care home. There were no visual, auditory or olfactory clues as to his presence but there was not the faintest doubt that he lived there or was visiting. Quite soon after this he came to work at my present care home, which indicates to me that he would suddenly have found himself thinking of me, unaware that I was standing outside the house, and found himself wanting to follow me. I know this is probably true because ever since I met him he has wanted to be friends but I strongly suspect that his motive is sexual, which I cannot tolerate, since I am totally heterosexual.

❖ 7. The quantification of metaphysics

Solid matter, as is known, is transparent to PP but, on thinking about it, one becomes aware that this may well be the case when the PP is in its usual state of rarefaction or tenuity, but since it is capable of existing at densities approaching that of flesh, there must be, for any given substance, a specific PP density at which that particular substance becomes opaque to it. (graph 4). As already written, I have a small 'umbrella PK demonstrator' in my room which I occasionally try to make rotate. (diags. 10, 11) and photo(fig 11). It used to move almost whenever I tried it, but since I put a small, clear-glass jam jar around it to prevent draught and breathing interference, it has refused to move, although I can still get half-hearted responses without the glass in place. I had assumed that its disinclination to co-operate was another example of the 'nonreplication effect' but now see that it is probable, that the 'Specific permeability', to create a new term, of the glass of which the jar was made, was very low compared to that of other substances. I should say here, to reassure the reader, that on the occasions when it did rotate, and once quite dramatically, for five minutes, I had made quite sure that neither draughts nor breathing could affect its motion. The glass jar was put there simply to reassure a sceptical friend, and in theory should have made no difference. Following on from this it becomes clear that every substance must have its own 'Specific Permeability', for a given standard tenuity of psychoplasma which could be, conveniently, basal density (measured in 'Rhine's' perhaps) or a more convenient level as on graph 4.The actual process of making this measurement would involve finding a place where one ccould be sure that PP density was at the basal level, perhaps in a desert or other isolated place, to calibrate the instrument. This takes me back to the occasions when I have witnessed the passage of solid matter through other solid matter i.e. the girl in my bed, flying off through the wall near my bedhead, the two young, brown skinned lads who appeared in my room a few weeks ago, only to disappear two seconds later, and on the next day, equally crazy, the guitar which I had left downstairs a few minutes earlier, impinging unexpectedly on my eyes. It stood there, leaning against the bookshelves and to do that it must have passed through four doors or several walls and ceilings. I am one hundred per cent certain that the guitar had passed through the intervening obstructions, walls, floors or ceilings without conscious human aid. Considering these last few sentences it becomes apparent that the concept of 'specific permeability' is insufficient to cover the observed phenomena. I have decided, however, to leave my description of the phenomena and the concept in place as evidence of my thinking on the subject so far, in order to save duplication of thought by subsequent workers on the theory.

It cannot be denied, however, that the phenomena do occur and the motion involved may take place in many ways and be due to many differing circumstances. If one considers all the possible ways in which an object can move or be moved from one place to another without the conscious intervention of a human agent one finds:

- It may float gently through the air to be deposited at its new location.
- It may disappear at place A to reappear instantly at place B with no intervening lapse of time. This would violate Einstein's theory of relativity but would agree with recent work in physics connected with Bell's theorem and the EPR paradox. (See chap.5 item 13).

- If it passes through intervening obstructions it may do so by oozing, osmosing or by percolating through, in a gaseous state.
- It may use a combination of these modes of transmission

❖ 8. The vacuum wheel psychoplasma detector

As mentioned earlier, I have invented a device which I call the 'vacuum wheel PP detector', which should make it possible to detect and confirm the presence of PP and/or BP. I do not possess the facilities to construct one myself but I shall describe its details so that interested parties can make one of their own. (diagrams.7,8,9.)

Essential components :
A drum or short cylinder (g) sealed at both ends and made of perspex or a similar material. It should contain a hard vacuum.(j) An axle (h) with bearings (b, f) on which it can rotate. A paddle (d) made of some flexible material, (rubber or something of that nature).bber, flexible plastic etc.) attached at point (k) to hub (e) of cylinder.
A motor or belt-drive system, belt to go round hub (e).
A frame or escutcheon (i) to carry wheel or motor etc. The recommended approximate dimensions are as follows:-
The perspex used to make the flat, circular faces of the drum or cylinder should be at least 15 mm thick to allow for atmospheric pressure, and the circumferential Perspex should be 10 mm thick. (the total atmospheric pressure on the circular face will be 1,610 lbs.).

Mode of operation.
The vacuum wheel contains no air but should be full of PP. If the wheel is set in motion and the paddle bends (m), this will indicate its presence, and that it has mass. The bending will occur as a consequence of the drag caused by the contained BP/PP. The vacuum is essential since if the cylinder contained air we would have no means of telling whether the drag was a result of air resistance or BP/PP resistance. If the paddle stays straight, or returns to its straight position after an initial bending, it will indicate that BP and PP do not have mass or viscosity in the normal way as described by physics, although it could indicate that the equipment was not sensitive enough. I prefer the latter because I know that both

substances exist. The experiment will not be conclusive if perspex is opaque to BP/PP but I think this is very unlikely.

It will, of course, be impossible to see the paddle's position if it revolves at more than two or three revolutions per second, so it will be necessary to use a stroboscope to 'stop' the revolving wheel. A stroboscope emits short, intense flashes of light whose frequency or 'repetition rate' can be continuously adjusted from a few flashes per second to one or two thousand. If our wheel is now rotating at 100 revs. per second the paddle will be quite invisible, but now turn on the stroboscope, let it illuminate the wheel and adjust its repetition rate to synchronise with the wheel's speed of rotation, and the wheel will appear to be stationary so that the state of the paddle, straight or bent, will become visible. It is a fact, of course, that the paddle itself will also be permeated by BP/PP, but I think that if the speed of rotation were high enough, this effect would be allowed for by BP/PP 'drag' on the paddle.

Unfortunately, since designing this device, I have discovered the reason for the unreliability of experimentation in psi, and I cannot say for sure that the device will work properly every time, but it need only work once to prove that there is some thing, i.e. PP, in the vacuum. The vacuum wheel detector could be very useful quite apart from its function as a detector, since it could be used to measure density of PP as well as detecting its presence.

If it fails the experiment could be done with glass, which I am fairly sure is transparent to PP if the PP is sufficiently tenuous. If this too fails the instrument could be made of some other substance, metal, plastic etc which would definitely be transparent to PP, leaving the wall of the drum (c) only which could be made of a transparent material, so that one could observe results.ERROR! I see now that this expt. would not work since the materiality of BP would cause it to be extracted along with the air to leave a vacuum.

❖ 9. The umbrella PK demonstrator (ref. 9c)
Take a two inch square of copy paper, fold it diagonally NE to SW. Next flatten it out and fold NW to SE. Flatten out and fold vertically N to S and lastly fold E to W. Now pinch it horizontally and vertically to produce a form which looks like the ridges of four rooves which join to form a cross shape.

Next take a sewing needle and embed the blunt end into a little blob of Blutack which is stuck onto a three inch square of card and balance the umbrella on the needle point. The indicator should be extremely sensitive and will rotate, under the influence or pressure of PP, if one simply observes it from a foot or two away. If it doesn't move the hands should be brought nearer or even cupped round it as though shielding the flame of a match. It is possible that the experiment will work once or twice and then will stop producing positive results. This has happened with my demonstrator and is an instance of the notorious 'non-replication effect' which is the plague of parapsychological experimentation, an explanation of which follows shortly. As written earlier the energy which is responsible for psychokinetic effects is primarily metaphysical - not kinetic - though at some point between the mental event and the physical effect, it must become kinetic. To be more specific, the metaphysical energy under discussion is essentially sexual since it is associated with the GW and the personal unconscious. This fact was nicely demonstrated on Monday 02/05/11 when I showed my umbrella PK indicator to a worker in the care home in which I am presently living. I cupped the indicator between my hands as though shielding a match from the wind, taking care not to touch it. Nothing. Not the slightest movement. I then let my mind dwell on a certain girl with whom I was in correspondence at that time. The indicator immediately began to rotate thus confirming that the energy was basically sexual. It didn't rotate, of course, if I thought of a male friend. This is a simple, straightforward experiment which anyone can do – but the mischievous 'non-replication effect' is at all times lurking in the shadows. My own indicator has steadfastly refused to show the slightest movement for several months now, with or without the glass jar in place.

❖ 10. More on the difficulty in replication of experiments in psi.
As explained earlier I believe that the god within is the fundamental active force in all psi phenomena. It cannot be pinned down by research for examination, analysis and classification, and will mischievously obstruct all attempts to do so; that is its nature and that is the reason why experiments in psi are so difficult to reliably zeplicate. It will, however, be perfectly happy to co-operate in applying psi to practical uses or useful technology, e.g.

dowsing for oil, water etc., tracing missing people by the use of a psychic or getting in touch with the afterlife through the offices of a medium and similar applications. (Ref.11d) It also supplies the metaphysical (basically sexual) energy that is needed to fuel the totality of human feelings, feelings of joy, hope, optimism, fear, uncertainty, delight, anger, sadness, love and satisfaction, which are metaphysical things and as such, not directly measurable.

The central fact is that psi can become perfectly replicable and predictable only when it is being used for some practical, useful purpose, and for this kind of technology, for that is what it is, (and why not develop it?),parapsychology will need to develop specific software, algorithms and even hardware; special techniques and procedures to perform particular tasks in society.

* 11. An example of Mutually beneficial feedback between applied science and pure research.

I have designed a procedure, or algorithm which should make it possible to accurately locate hidden objects or missing people, which is effectively

Fig.11. A photograph of my own psychokinesis demonstrator.

an original piece of psi-based technology, and it should make it possible for anyone to greatly amplify and apply the low-level powers that we all possess in some measure. I have done a trial test on the procedure which, though not entirely successful, gave startling results which suggested strongly that it would do exactly what it was designed to do. I am designing a 'hardware' electronic device based on it which I call 'Predictor' because I intend taking out a patent on it.

The technique should make it possible to circumvent the replication problem encountered in laboratory research, by removing it from an experimental context and using the procedures involved (the 'software') to do a useful job, providing one hundred percent accuracy in fulfillment of practical tasks and incidentally, as a 'spin-off', capable of being seen as a series of experiments, demonstrating perfect accuracy and replication in as many 'experiments' as desired.

I would like to describe the procedure, which, it should be noted, was suggested by the theft from my studio of three items the whereabouts of which I hope to track down with its use. Unfortunately I can't do this yet, because I would be giving away an idea which may enable me to make a great deal of money. When I have the patent its manner of working will automatically become public property.

Fig.12 A stares at B causing him to turn around.
(PP gathers along line of sight).

While we are on this subject of the impish nature of the GW we might as well take note of the fact that on the most basic level of all, it is obvious that the origin of life's manifest vitality must remain mysterious in its essential mischief. If it were not so life would be a dreary business; everything would be cut and dried, served up the same each and every day, rolled out grey and packed in identical boxes. Variety (and unexpectedness) is the spice of life as the old adage proclaims. The spark of life must remain elusive. The second that it doesn't it becomes lifeless. The basis of living things must remain impish, mischievous, unpredictable and essentially mysterious.

SUMMING UP

The marriage of physics and metaphysics So here we are at last.. Eileen Garrett describes objects in her boss's office 2,000 miles away, Uri Geller earns millions dowsing for oil companies, psychics dangle pendulums over maps to locate objects or people thousands of miles away; armies of ghostly Roman soldiers march through the damp and mossy walls of a cellar,

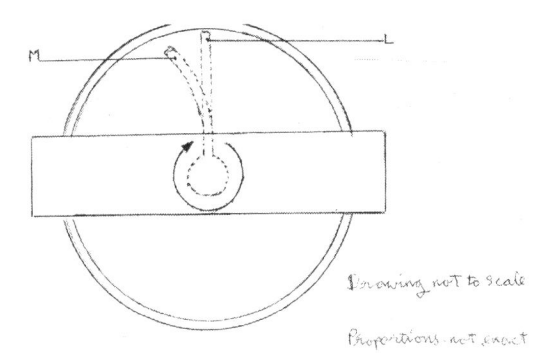

Diag. 7. Plan view of vacuum-wheel BP/PP psychoplasma demonstrator.

William Blake foretells the hanging twelve years in the future, of the man who should have been his drawing master, people leave their bodies at n ight to go 'astral travelling'. Psycho-kinesis, teleportation, levitation, poltergeists, precognition

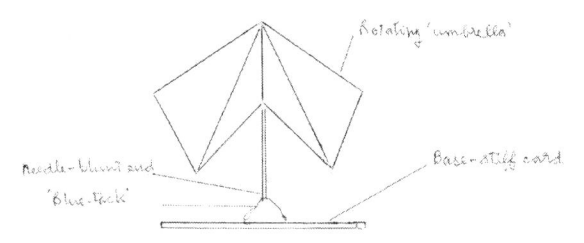

Diag.10. Umbrella psychokinesis demonstrator – side elevation

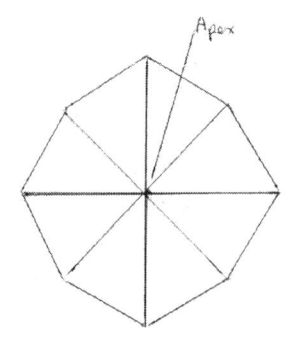

Diag. 11. Umbrella psychokinesis demonstrator – plan view

hauntings.......when one considers the manifold and bewildering variety of paranormal phenomena, it is difficult to see something which could serve as a locus for the unifying theory which should be the objective of psi research. I would like to suggest that this 'something' is non other than the revelation that the mind itself, both conscious and unconscious is, in composition, a material which permeates the body, the surrounding space and any material withwhich it comes into contact. That is - a substance - the key to all my following discoveries. The indispensable god within is an essential part of this 'something'. The activities and operations of a divinity are, by definition, not

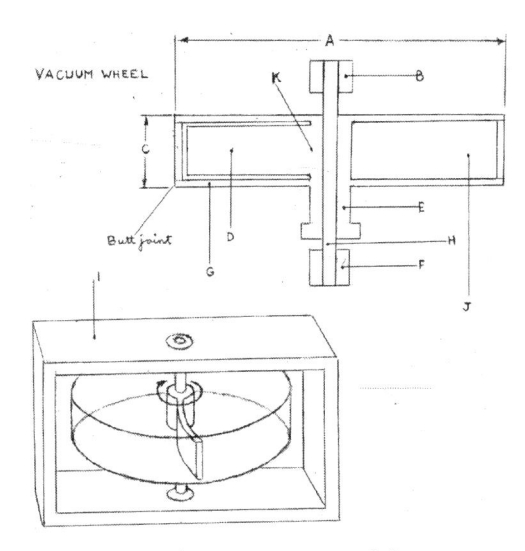

Diag. 8. and 9. Two views of the vacuum wheel BP/PP detector. (not to scale)

subject to the laws of physics; they are unknowable and inaccessible to science; they are ubiquitous in both the present and the afterlives and their sphere of operation is permeated by BP the medium on earth through which they work, and which extends probably to a height of 10 km. above ground and possibly the same below.

The apparent near-omnipotence of the god within, (limited only by its subjection to the will or spirit) is due to the material mind's ability to penetrate, permeate and intimately surround and infiltrate material things, close or distant so that virtually nothing is hidden from it.

Understanding and explaining the phenomenon of consciousness is now one of the major objectives of science. It is sometimes known as 'the hard problem', a name coined by David Chalmers, a dualist, who believes that there must be an extra phenomenal realm where consciousness can be found. It is called the 'hard problem' of consciousness, in opposition to 'the easy problem'- researching the physiological structure of the brain. Roger Penrose, the physicist who worked with Stephen Hawking on black holes and the big bang, postulates sub-neuronal entities called 'micro-tubules' as its origin. It is at this point, I suggest, that some degree of connectedness with the intelligences which must, I think, populate spacetime may occur. There is no doubt in my mind of the existence of the god within, but one is lead by this fact to speculate as to what happens to it at the death of the organism. It isobvious that apart from the gods within which energize the living organism, there exists a plurality of discarnate divinities, those associated with departed souls, and possibly others, whose doings are, of necessity, beyond the reach of science, and which come to our notice only when a new organism comes into being, or an otherwise inexplicable event occurs. It is clear from my own experiences that PP can interact with matter on a subatomic level (electronic bathroom scales, mobile phone, computer, pocket calculator etc. where, one presumes, electron flow is interfered with).

Physics then, is possibly on its way to a final state where the three forces have been unified, or beyond which no further progress can be made. The biggest picture of all, the unifying of physics and psi, if possible, is a long way off in the future and the field to be explored is vast, exciting, inspiring and when the exploration is complete, if ever, the world (and its outposts - worlds beyond worlds)will be transformed beyond recognition.

Since writing the above I have had many more even more astonishing experiences which I intend recounting details of in a sequel to this essay. One of the more extreme is that under certain circumstances (when I was evading duties which were made incumbent by the nature of the phenomena), I found myself actually affecting the weather, in such a way that a 'black' feeling grew up and a sudden wind whipped up which threatened my safety, by closing the doors of a rubbish dump wherein I was stowing rubbish. I knew in an instant what was happening since it corroborated certain similar happenings in my past, and was obliged to modify my behaviour since it was impinging badly on the comfort of one of the support workers in the carehome where I live. I was, essentially, over-reacting to an imagined attack by this person and was, in effect, committing an evil act. The imminent entrapment in the rubbish dump was a kind of punishment meted out by the various gods within (or without) who were involved.

I do, actually, see how crazy and unlikely this must appear to the reader, especially in the light of my sojourn(s) in my local mental hospital. These, however, actually caused a lot of the phenomena which themselves I was able to treat simply as 'more grist to the mill', despite their often uncomfortable, and sometimes very painful nature.

I would like to end this essay, or rather, book, for such it has become, by quoting a passage from a poem by William Wordsworth, which expresses its essence in a few eloquent lines.

A Quotation from
'Lines Composed a Few Miles Above Tintern Abbey'

For I have learned.
To look on Nature, not as in the hour
Of thoughtless youth, but hearing oftentimes
The still, sad music of humanity,
Not harsh nor grating, though of ample power
To chasten and subdue. And I have felt
A presence that disturbs me with the joy
Of elevated thoughts; a sense sublime
Of something far more deeply interfused,
Whose dwelling is the light of setting suns
And the round oceans and the living air,
And the blue sky, and in the mind of man;
A motion and a spirit that impels
All thinking things, all objects of all thought,
And rolls through all things

* * *

DIAGRAMS AND ILLUSTRATIONS

Diag. 1: The universe 5

Graph 1: Effects of altitude and locality on psychoplasma density. 9

Fig 1: The Kirlians' missing leaflobe 11

Fig. 2. The spirit in the heart and the god within 21

Fig. 3 The gods within reach out to join with other gods in other people and higher animals. 26

Graph 2: The density of PP within and between groups of people versus distance. 36

Diag. 2. Plan view of a large material conscious mind (shown unconnected with other minds). Personal Unconscious – not shown – would lie behind the conscious one 39

Fig. 4 Conscious and unconscious telepathy. 40

Fig. 5. The tesseract or hypercube 46

Fig. 6. Distribution of memories in the human body. 63

Fig.7. Conscious and unconscious minds joined and contained within a discontinuous boundary (effect of neuroleptic drug quietiapine ingested by A) 81

Diag. 3. The planet showing psychosphere, collective conscious and universal unconscious 88

Fig. 8. Temporarily detached part of material mind rejoins main mass 95

Fig. 9. Most mental activity takes place outside the head 111

Graph 3. Density of PP relative to differing parts of the landscape and flora and fauna. 132

Diag. 13. World line for two ships on a collision course C, a horse race A and B, winner seen in dream, and D two other ships and a premonition and the pyramids whose spatial position does not change. 136

Diag. 4. Retro-time –travel through psychoplasma = temporal teleportation 141

Diag. 5 Group minds. Contour defining regions of indicated PP density. All points on any given line will have same density. ..167

Fig. 10. Plan view of collective unconscious and collective conscious at ground level. 168

Fig.11. A photograph of my own psychokinesis demonstrator. 200

Fig.12 A stares at B causing him to turn around. 201

Diag. 7. Plan view of vacuum-wheel BP/PP psychoplasma demonstrator. 203

Diag.10. Umbrella psychokinesis demonstrator – side elevation 203

Diag. 11. Umbrella psychokinesis demonstrator – plan view 204

Diag. 8. and 9. Two views of the vacuum wheel BP/PP detector. (not to scale) 204

* * *

GLOSSARY

AL   The afterlife

BP   Bioplasma, a substance secreted, I think, in the cells of the body and in all other living matter.

BPE   The bioplasmatic envelope which connects all living matter, including that in the afterlife and is the medium of unconscious telepathy and many other occurrences of psi.

CC   The collective conscious. The CC is the conscious complement of Jung's collective unconscious and its constituents are BP and 'spirit' which combine in the body to produce psychoplasma – the substance of the conscious mind.

CU   the collective unconscious. As described by C. G. Jung but with the additional characteristic of being material, that is, made of bioplasma – the substance of the unconscious mind.

DI   The duality of imagination. Imagination has two aspects, the material one composed of psychoplasma and the Platonic one through which one can gain access to a purely spiritual world of forms. DI is simply the fact that since the mind itself is material any event which occurs in it is also material, so that it becomes difficult to 'imagine' an object or entity without its necessarily possessing substantiality. This does not apply when one imagines, or tries to imagine, ideal forms i.e., perfect geometrical figures, or considers numbers, which are, of course, exact. Truly perfect forms, entities and concepts cannot be found in the imperfect world of matter – they belong to the world of spirit, which, as shown in Chapter 2, item 4 does exist. (See also, Chapter 2. item 5) This is, basically, a platonic world view.

GW	The god within is essentially the origin or necessary agent in all parapsychological phenomena. It is singular in each higher animal and exists in all other forms of life as part of a world wide amorphous and mischievous, though not malicious presence, which in earlier times was known as 'God'.
IOTS	The individuality of the spirit. There exists only one of any particular spirit or identity and it cannot be divided in such a way that that it dwells within more than one body at any given time.
OBE	Out of the body experience. The subject has the sensation of rising from his/her body and ascending to ceiling height, and into other rooms in the building. It is believed (by me, and other investigators) that the ascending component is the spirit of the subject, and in my opinion the spirit is the basis of consciousness.
PC	Personal conscious mind, the motivating component of which is the spirit in the heart.
PK	Psychokinesis. The capacity to cause movement of matter without recourse to physical contact.
PL	Present life.
POF	The Primacy of Feeling. A general principle which states that if feeling and logic conflict logic must give way unless the feeling can be adjusted to accommodate the logic, thus achieving a state of balance. If not feeling must prevail and ignoring it will usually have regrettable consequences. This principle applies on a personal and a universal scale.
PP	Psychoplasma. This is the substance formed by the combination of bioplasma and spirit and it is the stuff which connects minds together. It is the physical basis of telepathy. I believe that it is saturated with information much of which is in a state of flux. I think it must also contain more or less solidified fields or constituents which are related to memory.
BF	Biological field. The BF is essentially a synonym for the conjunct conscious and unconscious material minds, which emphasizes its short-range physical 'stickiness', and its power to bind adjacent people together.

PS	Psychosphere. The sum of the UU and the UC or the totality of psi phenomena on the planet including the afterlife.
PU	Personal unconscious.
RV	Remote viewing. The power to see a distant place or object through the eyes of a 'beacon' or distant person who is observing it.
S	Spirit. The substance, (if it is a substance) which combines with bioplasma to make psychoplasma, or a synonym for the spirit in the heart which goes, with the mind, to the afterlife at the death of the organism.
SH	Spirit in the heart. The 'I' or essential self which is the basis or motivating agent in the conscious mind.
TMM	The theory of material mind.
UC	Universal Conscious. The sum of the Collective Conscious of the present life, and the collective conscious of the afterlife. The medium by which psychics get telepathic access to the AL. It is the conscious counterpart to the UU but is much more tenuous. I have gradually come to believe that it is identical with the CC which is shared by the afterlife and the present one.
UMSP	Unconscious mind suppression phenomena. These are commonly known as 'poltergeists' meaning mischievous spirits, but they are in fact results of the god within, in the unconscious, diverting suppressed metaphysical, basically sexual, energy into the environment where it appears as kinetic energy.
UU	Universal unconscious. The UU is the single most powerful component in the psychosphere, and is synonymous with the BPE.
TFITK	To feel is to know. This means that in a deceptive or unreliable exchange with a particular person, or a situation where the logic is uncertain the feeling rather than the words, conveys the truth.
SP	Perception of a remembered feeling

* * *

AN AID TO ASSIMILATION OF CONCEPTS AND IDEAS

For the sake of clarity and as an aid to understanding and arranging the concepts and information in the reader's own mind, I am including a simple list of:

(A) Facts which I know to be true, but which may remain to be proven
(B) Beliefs which I am fairly sure are true, and
(C) Speculations

FACTS

1. The conscious mind is composed of a material which I call psycho-plasma and is not confined to the head or body but extends into the surrounding space. The vacuum-wheel PP detector as described in chapter 8 is designed to prove this but an actual working example has not yet been constructed.
2. The basic motivating force of the unconscious mind is a divinity, the god within, which resides in the gut.
3. There is a spirit in the heart of every human being, which animates the body and goes to the afterlife at the death of the organism. It is also the motivating force of the conscious mind, and through that entity, much of the body in general.
4. The universal unconscious (UU) is an amalgamation of the collective unconscious of the present life and that of the afterlife, and it is at the same time identical with the BPE which is shaped

and informed by the sum total of all gods within, that is, those in the present life and those in the afterlife.
5. The gods within communicate via a material medium which I call bio-plasma (BP) which constitutes the BPE, encircles the earth, and is the material embodiment of Jung's collective unconscious, and my universal unconscious.
6. The collective conscious is made of psychoplasma (PP) and girdles the Earth. It is the sum of all conscious minds, plus ubiquitous 'free' PP which links all conscious minds together. It is the medium of conscious telepathy, and the matrix of much, but not all psi phenomena; e.g. poltergeist phenomena are the result of bio-plasmatic activity which is unconscious.
7. Imagination is dual. That is, one aspect, e.g. logic and mathematics, operates in an ideal, Platonic world, and the other in a more abundant and frequently encountered world of the material i.e. psychoplasma.
8. Poltergeist phenomena are not spirits but a consequence of the conscious mind, forced to conform with constraining social circumstances, suppressing the unconscious mind and the GW, which then, unable to express its energy normally, directs BP to perform acts of unconscious PK.
9. Memories are stored not only as traces in the brain, but are located throughout the body, and even outside it.
10. The expression to 'Feel is to Know' (Already defined in glossary).
11. The Primacy of Feeling is a universal conditioning factor in human affairs and it means that in any argument or social system the importance of the feeling is equal to or greater than that of the associated words or logic (discrete), which are often open to abuse and may allow subtle errors to creep in, whereas feeling (continuous), is real and indubitable. An illustration of this is an interesting type of sophism in philosophy called 'sorites', in which a series of at first truthful statements gradually leads, through a perfectly respectable logical argument, to an absurd conclusion. The basic premise is that adding one hair cannot make a difference as to whether a man is bald. An example is the man who had only one hair on his head. He is, of course,

considered to be quite bald. If he were to grow another hair he would still be bald, and if he grew another to make three he would still be bald. This process can be extended so that at one hundred hairs, added one by one, he would still be bald and so on until he had a rich luxurious thatch and was still bald. This is an example of logic and feeling coming into conflict. Which does one trust? At some point, if this were being demonstrated to a student, let's say, he would carefully follow the argument, but at some point the feeling of trusting belief would change to one of amused enlightenment and he would laugh and say "yes, I get the idea". He obviously couldn't let logic have its way and convince him that the hairy man was bald. It would feel utterly wrong. This illustration is similar to Thomas Aquinas in the 13th century discussing the question of how many angels could dance on the head of a pin. (and physicists lead by the nose, or rather, mathematics, [logic], into believing the theory of the 'big bang' which I, along with others, believe to be untrue).

12. The bioplasmatic envelope (BPE) envelopes and connects all life that exists or ever did exist.
13. Psycho-plasmatic bonds grow between frequently or permanently contiguous human, animal or vegetal life.
14. In one of my many contacts with the afterlife I found myself in telepathic contact with a hominid which had died 7,000,000 years ago. My contacts with the afterlife are all examples of 'to feel is to know'.
15. The god within operates outside the laws of physics, and transcends space and time.
16. Feelings can be visceral, emotional or mental.
17. Psychoplasma is responsible for the warm feeling of friendship that one gets from one's fellow-humans, and even one's pets. It flows from one body into another and its absence is the cold feeling of loneliness.
18. In matters of perception, at bottom it is the spirit in the heart which perceives.

BELIEFS.

1. Teleportation of human beings, other organisms and inanimate matter is possible and should be fairly straight-forward, given time.
2. Out of the body experiences provide a clue to the origin of consciousness, whose basis is not physical, but metaphysical. Although consciousness is perceived, felt, in the head, its origin is not to be found in the brain, but in the spirit in the heart, the essential self.
3. The mind (head plus heart) is stronger than the body. The mind can exist as a living presence without the body e.g. the OBE, but the body without the mind is dead matter. That is, the body needs the mind to survive, but the mind can survive without the body. This argument is slightly weakened by the presumed fact that the GW remains immured in the body during an OBE so that the mind and body are not completely separated at such times. This qualification, however, is negated by the fact (not actually proved as yet but known by me to be true) that the afterlife exists, and with it the departed spirits of once living organisms; also it is augmented by the acts of the self-immolating buddhist monks in the sixties protesting about the Vietnam war. (chap. 2 item 5b)
4. The individual conscious mind is finite, but the unconscious is infinite and transcends space and time. It is responsible for intuition and creativity.
5. The god within is androgynous.
6. The current and ongoing emergence of hitherto suppressed pagan beliefs is a vindication of the theory's claim that intelligent agencies are operating within, around and between the life-forms which populate the earth.
7. Science has arrived at a juncture where it will be under increasing pressure to readmit metaphysical truths into its thinking, and will be forced to accept that its knowledge will be forever incomplete (ref. 4b).

8. At bottom the universe is not only probabilistic but is founded in mystery and operates by, what to us,. can only seem like magic. There are divine and spiritual intelligences reappearing in Nature, after being thrust into the background by the successes of physical science.
9. Eternal youth. (The alchemists' elixir of life). I believe that through judicious deployment of the powers bestowed by the theory, one can double or even treble one's life-span, preserving at the same time, the 'bloom of youth'. The essential ingredients or qualities necessary to realise this objective, is a firm belief or, better still, concrete knowledge and understanding, that it is possible, and it involves, unlike orthodox science, believing *despite* negative evidence, which normally would weaken an argument. The process is very similar in some respects, to the breaking of the barrier of the ' four minute mile' which, since accurate measurement of time had been possible, had not been broken. As a consequence of this it was generally believed that the 'four minute mile' was impossible. Then, in 1954, Roger Bannister ran it in 3 minutes 59.4 seconds. Soon after the barrier was broken again by Christopher Chataway and others which proved the barrier to be psychological. I believe that the current life expectancy in any society is determined by similar factors in addition to medical advances.

It is an accepted fact that feeling and appearance are linked to the stresses encountered in life, and that if a period of calm, peaceful prosperity follows one of violent stress and anxiety, the subject gradually becomes healthier and younger in appearance. These determinants operate in the 'elixir of life' context too, of course, but the theory brings a lot more information and wider possibilities into the equation, namely, the god within which supplies unlimited funds of good, youthful feeling, as long as the subject remains open and retains his belief in the possibility.

Ultimately the essential ingredients for attaining a long, protracted life span, are, as always, (not counting taking good care of oneself), pure drive informed by a strong belief, plus a clear understanding, that it is possible, and an occupation or set of objectives, hobbies, whatever, that one does purely for love; this includes the elevation of the soul or spirit that the right kind of music can bring, be it pop, rock, Beethoven, Bach, jazz or folk.

The proposed method for starting the process through exercises is as follows:

The method relies for effect on the actual materiality of mind and subsequent materiality of imagination.

(a) First step. Remember the feelings of being young. Summon up memories of good times, old friends, the music that one liked in those days, places, pubs, clubs etc.. These memories will probably occur at head level.
(b) Let these memories 'percolate' downwards through one's neck and shoulders, into one's heart, meanwhile walking or freely moving about. This may be quite difficult at first but should get easier.

when I'd got a job as a technician in the sculpture school at Hammersmith School of art, as a precursor to applying for a place as a student. I remember sneaking regularly into the life drawing class to build up a folder of drawings, and remember the teacher, Mr. Folkard, admiring my work and asking who I'd been with before, my former teacher that is, and my explanation that he was my first one. At this point (in my expt.) I realised that this whole business of recalling the past was unnecessary because as one proceeds through life, preserving a positive outlook, memories – cheerful, happy, youthful ones will naturally recurr, and, moreover if one has properly understood the role of the GW they will be accompanied by strong, solid feelings. Again, what this actually means, is that if one has properly understood the role of the GW all one need do is get on with one's life and gradual enlightenment will follow; that is, one will find oneself from time to time with the magical feeling of youth described earlier and if one continues to believe it will keep happening as it did - and does – with me.

It would, perhaps, be better had I chosen music associated with my thirties or forties, thus leading to more appropriate feelings, and a more probable re-acquisition, over time, of a more youthful appearance.

However, the process is as illustrated whatever one's target age. It is essential that one does not paint these memories in nostalgic, melancholy tones for they are the raw material of gradual rebirth - mental rebirth - and the mind, as is well-known, strongly affects the body, and ultimately, as is shown in chapter 2 item 5 (the dual nature of reality) is a stronger and more independent entity.

This causes one, through the materiality of imagination, to *feel* in one's heart and body (not just remember in the head) the feelings which cause one's mind and body to become young again, (provided that one hasn't let things go too far), or to remain young. This also entails, I think, automatic reparation of cell atrophy by the subatomic effects of the GW although this would not be the main mechanism. (The GW and the spirit are constants, they don't change essentially, whereas the body and conscious mind are variables).

This option of retaining or regaining one's youth need not necessarily involve exercises, although exercise is good, it is sufficient to know and understand the mechanism's involved so that one can conduct one's life in such a way as to delay or prevent the process of becoming old. I myself in normal life, whatever I am doing, am always very aware that certain activities or circumstances make me feel younger and others older, so I conduct my life accordingly. I had an excellent experience only yesterday while typing; I suddenly felt that I had become the person that I was at age twenty. Not just a feeling of being well and full of energy but a feeling of light, cool youthfulness which filled my heart. I simply felt a magical sensation of being myself again, that person age twenty, whom I had forgotten. I felt, as I always do at such times, that I'd 'woken up'. It would seem that growing old is simply a long process of falling asleep and forgetting the young person that one once was.

10. I believe that by the year 2150 or thereabouts, it will be possible to travel by teleportation from london to New York or between any of the world's capitals, in a journey-time of minutes, seconds or even, possibly, no time at all.

SPECULATIONS AND EXTRAPOLATIONS.

- It may be possible within a century or two, to teleport from place to place on the earth's surface, resurrect the departed, and, from a position in the AL, travel into the remote past. It may also be possible to time-travel back to one's youth. It may also be possible to travel into the future; this, clearly, was possible for the Greek philosophers and the hominid but possibly only because I had accidentally summoned them. All this could obviously have a beneficial and increasing effect in reducing, stopping or even reversing global warming.
- It may be possible within thirty years to cure cancer and other conditions, perhaps all conditions, by subatomic psychoplasmatic intervention.
- TMM Bids fair to be the alchemists' sought-after elixir of life, the key to eternal, or at least extended, life.
- It is probable that in the years to come the theory of material mind will have diversified into many related divisions so that there will be researchers, specialists, analysing the nature of BP and PP, investigating the afterlife, specialists in PK, precognition, remote viewing, teleportation, time travel, specialists exploring the physiology of the interface between body and spirit, and body and god within. These will all be subjects in their own right. (Unless the god within objects and deploys its incorrigible mischief!)).
- If aliens have visited Earth as recently claimed by a certain important person in touch with UK officials in charge of military intelligence, (BBCbreakfast@bbc.co.uk.5 July 10), and if they really could read his mind and knew all the secrets of our intelligence, they must certainly have known the theory of material mind and developed a highly sophisticated technology based on it. They would also be able to travel much faster than the speed of light, either by physical means, the hypothetical 'wormhole', or some technology made possible by the space and time transcending powers of the god within, in a 'non-local' universe. (See chap. 5 item 13).

- If aliens really do exist and have visited Earth, they must incorporate one or more metaphysical components in their being.
- If there were no biological bodies there could be no bioplasma and therefore no psychoplasma, since BP is (I believe) secreted in the cells of the body. There would be only the god within itself and the spirit. Perhaps this would be the scenario before 'the big bang' if such an occurrence ever did happen? No living bodies – no matter, or physical energy – only metaphysical energy (GW) and psychic energy (spirit), which itself is metaphysical though it is more convenient to think of it as the essence of the person or organism.
- The basic mystery of quantum mechanics resulting from the Heisenberg indeterminacy principle, which states that the exact position and velocity of a subatomic particle at any particular instant cannot both be known, may be exactly the same thing as the mystery of the god within which is equally unknowable. This could then be the region where physics and metaphysics overlap and unite matter and spirit. (Nobody knows, or even can know, exactly how quantum mechanics works; its roots are hidden, just as nobody can know how the god within works. It will be forever a mystery). See Addenda 107 and 108
- Einstein could not accept quantum mechanics; the idea that the universe was ruled by chance and probability worried him even though the maths indicated that to be the case. Physicists themselves did not understand the mechanisms involved and had to simply live and be content with the knowledge that if they stuck to large numbers the behaviour of particles could be predicted, and matter manipulated. This gap in causation in quantum theory is perfectly filled by the presence of the ubiquitous god within and, possibly, the gods without which, in addition to the GW may be real, dynamic presences, operating, to all intents and purposes, through what can only seem like magic.

* * *

REFERENCES

1. Extracts from the writings of C. G. Jung.

   (a) the view that phenomena, rather than being a-causal, are due to magical influence (Jung 1952-501-2). What early theories suggest to him instead, is that there may be a meaning that does not depend on human subjectivity but is "transcendental" or "self-subsistent" – a meaning which is 'a priori' in relation to human consciousness and exists outside man. (My comment on the above is that the actions of the gods within are essentially, necessarily magical and mysterious, meaning not comprehensible to orthodox physics nor even parapsychology itself, and are beyond scientific investigation at present or at any time in the future.- Author).

   (b) as though space and time did not exist in themselves but were only "postulated" by the conscious mind (Jung 1952:435). Knowledge of events at a distance or in the future is possible because within the unconscious all events exist timelessly and spacelessly. (This assertion fits in very well with the idea of the afterlife and the apparent ease with which one can contact a given person on the same day, as an old man, or a young one. See author's psi experiences, chapter 8 item 6).

   (c) Jung was staying in what was reputed to be a haunted cottage and was being troubled at night by a cacophony of strange noises and other nasty goings-on. A quotation from Jung: 'the weekend was so unbearable that I asked my host to give me another room. This is what had happened. It was a beautiful night, with no wind. In the room there were rustlings, creakings, and bangings: From outside blows rained on the wall. I had the feeling there

was something near me. There beside me, on the pillow, I saw the head of an old woman, and the right eye, wide open, glared at me, half of the face was missing below the eye. The sight of it was so unexpected that I leapt out of bed, lit the candle and spent ……. the night in an armchair.'
Jung on synchronicity and the paranormal by Roderick Main page 66. Princeton ISBN 0-69105837-7

(d) When I visited Freud in Vienna1909 I asked him what he thought of these matters (parapsychology) he rejected this whole complex of questions and did so in terms of so shallow a positivism that I had difficulty in checking the sharp retort on the tip of my tongue. While Freud was going on I had a curious sensation. It was as if my diaphragm were iron and was becoming red- hot –a glowing vault. At that moment there was such a loud report in the bookcase, next to us, that we both started up in alarm, fearing the thing was going to topple over I said to Freud, "there, that is an example of catalytic exteriorization phenomena". "Oh come," he exclaimed, "that is sheer bosh". "It is not" I replied, "you are mistaken, Herr Professor, and I now predict that there will be another loud report!" Sure enough, no sooner had I said the words than the same detonation went off in the bookcase..C. G. Jung. Memories, dreams, reflections, page 178. Fontana ISBN 0-00-642519-4

2. Extract from 'Mysticism and logic' by bertrand Russell.
Physics tells us that certain electromagnetic waves start from the sun and reach our eyes in about eight minutes at the end of this purely physical series, by some **odd miracle** (my italics -the operations of the spirit [the seat of perception] in the heart, in tandem with the head, which, I believe, together are the origin of consciousness –Author) comes the experience which we call "seeing the sun", and it is such experiences which form the whole and sole basis for our belief in the optic nerve, the rods and cones, the ninety-three million miles, the electromagnetic waves and the sun itself. physics has been lead into the construction of the causal chain inwhich our seeing is the last link and the immediate object which we see cannot be regarded as that initial cause which we believe to

be ninety-three million miles away, and which we are inclined to see as the "real" sun.(From Mysticism and Logic by Bertrand Russell. Page 105. Dover publications Inc. Reprinted in 2004 ISBN 0-486-43440-0).

3. Extract from 'The 21st Brain' by Steven Rose.
At least the intestines contain neurons and the general regions of the bowels are rich with hormone secreting organs. So today's "gut feelings" have something going for them. The metaphorical division between heart – or gut –and head, remains embedded in present-day neuroscience, with its insistence on dichotomizing between cognition and affect. (Author's note [if I may be so bold]). I think from what I've seen, felt, and reasoned, that the primary dichotomy is between the gut and the conscious mind (head + heart) and the secondary one is between head and heart).(From The Twenty-first Century Brain by Steven Rose. Page 194. Vintage Books. ISBN-099-42977-2)

4. Extracts from 'Arthur Koestler, the Homeless Mind' by David Cesarani.

   (a) Modern physics and parapsychology he asserted, were converging. It was only a question of time before the, as yet, mysterious substance which enabled thought to be transmitted, would be revealed, as surely as had the existence of the neutrino been uncovered..
   (b) Similarly in a broadcast for the American radio network NBC at this time, he concluded that, "both physicsand parapsychology point to aspects of reality beyond the reach of contemporary science, a coded message, written in invisible ink between the lines of a banal letter.(From 'Arthur Koestler, the Homeless Mind' by David Cesarani. Page 518 Published by William Heinemann Ltd. Reprinted by permission of The Random House Group Ltd. ISBN 0-0992-8967-9)

5. Extracts from 'Descarte's Error' by Antonio Damasio.

   (a) I suggest only that certain aspects of emotion and feeling are indispensable for rationality. At their best feelings point us in the direction, take us to the place in a decision-making process, where we can put the instruments of logic to good use.

(b) Emotion, feeling and biological regulation. All play a part in reason. The lowly orders of our organism are in the loop of high reason (note: my assertion that the oldest memories are stored in the intestines at the bottom of the torso, and others are stored all over the body Chap.3 b)

(c) The physiological operations that we call mind are derived from the structural ensemble rather than from the brain alone.

(d) he showed the subject emotionally charged stimulae, for instance, houses burning, people about to drown in floods he told me that he could sense how topics that once had evoked a strong emotion, no longer caused any reaction. This was astounding try to imagine yourself not feeling pleasure when you contemplate a painting that you love or hear a piece of music. Imagine yourself robbed of that possibility and yet aware of the intellectual contents of the visual or musical stimulus, and aware also that it did give you pleasure. We might summarize Elliot's predicament as 'to know but not to feel'.(at the centre of my own thesis lie the words, 'to feel is to know' and 'the primacy of feeling').

(e) His particular target is the dualism that splits 'mind' from 'brain' but his own solution does not stop at simply saying that conscious experiences come from brain states. His view is that minds are embedded (he prefers the term, embodied) not only in the brain but in all of the rest of the body.

(f) Perhaps the human mind is such that the solution to the problem can never be known, because of our limitations. Perhaps we should not talk of a problem at all, and speak instead of a mystery drawing on a distinction between questions that can be approached by science, and questions that are *likely to elude science forever.*(Author's italics)

(g) the cold-bloodedness of Elliot's reasoning prevented him from assigning values to different options and made his decision-making landscape hopelessly flat.

(h) I see feelings as having a privileged status. They are represented at neural levels, including the neuro-cortical, where they are the neuroanatomical and neurophysiological equals of whatever is appreciated but because of their inextricable ties to the body, they come first in development and *retain a primacy* that subtly pervades our mental life (author's italics) feelings have a say on how the brain and cogitation go about their business. Their influence is immense.

(i) Our sense of wonder should increase before the intricate mechanisms that make such magic possible (love, art etc.). Feelings form the base for the human soul or spirit. (which *is* an existing entity, after all, independent of, and controlling, the body.- Author)

(j) Contrary to traditional scientific opinion, feelings are just as cognitive as other percepts. (Hence, my dictum: 'to feel is to know'.- Author) From ' Descartes' Error'by Antonio Damasio. Vintage 2006 ISBN 978-0-099-50164-3

6. Extract from the writings of Peter Tompkins and Christopher Bird.
Plant Life'..The Secret Life of Plants by Peter Tompkins and Christopher Bird. Page 23-25. Penguin Books 1974

7. Extract from the writings of Stuart Holroyd.
In 1967 two Russians, V.S. Grischenko and Viktor Inyushin, reported research in which they coined the terms 'bioplasma' and 'bioplasmic body'.' Our experiments indicate', wrote Inyushin, 'that bioplasma consists of ions, free electrons and free protons, in other words, subatomic particles that exist independently of a nucleus'. Physical bodies, he went on to say, have 'biofields' constituted by bioplasmic particles that emanate from the body, forming a kind of envelope surrounding the physical body these emissions may be involved in telepathy and psychokinesis.
The Arkana Dictionary of New Perspectives, by Stuart Holroyd. Page 152 Arkana ISBN 0-14-019195-X

8. Extract from the writings of Gary Moring.
Because the role of the observer plays an important part in the study of the quantum world, there may be a close relationship between

the dynamic structure of the quantum universe and the nature of the unconscious. From The Idiot's Guide to Understanding Einstein by Gary Moring. Page 247. Alpha books. ISBN 0-02-863180-3

9. Extracts from the writings of Colin Wilson.

> (a) In the 1960s the psychiatrist Charles Tart studied a borderline schizophrenic girl whom he called Miss Z., who told him she'd been leaving her body ever since childhood. To test whether these experiences were dreams Tart told her to try an experiment: she was to write the numbers one to ten on several slips of paper, scramble them up, then choose one at random when her light was out and place it on the bedside table. If she had an out- of- the-body experience in the night she was to try to read the number (she claimed to be able to see in the dark during her OBEs). She tried this several times and found she always got the number right. So Tart decided to test her himself. The girl was wired up to machines in his laboratory and asked to try and read a five digit number which Tart had placed on a high shelf in the room next door. Miss Z. reported correctly that the number was 25132. Page 252.
>
> (b) Fred Hoyle and the chemist Laurence Henderson argued that the universe seems almost unreasonably suited to the existence of life. In the 1950s Hoyle was working out how the elements are created in the heart of the stars. He noted that, in order to create carbon, - the element essential for the creation of life – two helium nuclei have to collide, a contingency as unlikely as two billiard balls colliding on a billiard table the size of the Sahara desert. But when this has happened the new atom seems to attract a third helium atom to make carbon: no other element behaves in this way. Moreover if another helium atom hits the atom it produces oxygen, another element essential for life Hoyle came to the extraordinary conclusion that:
> "A commonsense interpretation of the facts suggests that a 'superintendent' has monkeyed with the physics – as well as chemistry and biology – and that there are no 'blind forces'

worth speaking about in Nature. I do not believe that any physicist who examined the evidence could fail to draw the inference that the laws of nuclear physics have been deliberately designed with regard to the consequences they produce inside stars". This is only one step away from saying that these laws have been designed to produce life.

(c) The umbrella PK demonstrator. I got this idea from Colin Wilson. From 'Beyond the Occult' by Colin Wilson. Pages 252 and 469. Watkins. 2008. ISBN 978-1-905857-69-2.

10. Extracts from the writings of J.B. Rhine.

(a) The question is merely, 'Is there anything extra-physical or spiritual in human personality?' The experimental answer is yes. There now is evidence that such an extra-physical factor exists in man. The soul hypothesis as defined has been established, but only as defined. Page 174.

(b) What has been found might be called a psychological soul. It is true that, as far as we have gone there is no conflict between this psychological soul and the common theological meaning of the term. The first step was essential, however modest. It has established a point that millennia of argumentation have failed to make. This beginning represents the turning *of three centuries of domination of our science of human nature by Physicalistic theory.* (Author's italics). It will eventually have the most revolutionary significance, though the full effect may be slow in being seen. The turning of tides is never sudden.

(c) Chapter 5. Across the barrier of time – precognition. The Reach of the Mind by J.B. Rhine. Penguin 1948 Page 175.

11. Extracts from 'Beyond Supernature' by Lyall Watson.

(a) Alan Gauld, a psychologist at the university of Nottingham, has made a survey and computer analysis of 500 cases (of poltergeist phenomena) he found that breakages and injury occurred in around 10% of his cases, without evidence of structural damage to the houses concerned. Animals were disturbed in 6% and another 16% involved assault – invisible pinches, blows, scratches and bites – on human beings. He could find no evidence that any of the breakages or injuries were the result of fraud or accident, and admitted to being impressed by cases such as a Canadian one of 1889 in which something near Quebec City " took money, spread excrement around, stole food, broke windows, caused fires, cut off hair, threw things, appeared in grotesque forms, developed a voice, swore, blasphemed, repented, became pious, blamed a witch, sang hymns, assumed the figure of an angel and before leaving said farewell to three children." The implication is that, at some level and in some way, consciously or unconsciously, either as a stimulus or as a response, *intelligence* is involved. (Author's italics) Page 192. (It seems appropriate here, to assure the reader that the god within alone would not commit assault of any kind or degree, and that, in cases like this, the guiding spirit which is obeyed without question, must itself be where the evil is to be found).

(b) Charles Honorton explored the effect of psychological variables on attempts to influence a random number generator, and discovered that scores only became significant when his subjects were working as a group. Individually, they failed. (p.202). (this is explained by the fact that the GW is the active agent here, and it awworks by means of infectious enthusiasm which is the god's most salient quality.

(c) BBC reporter Dick Tracy went to interview a couple in their "haunted" home. After a recording, he discovered there was nothing on the tape, but as soon as he went outside the interview played back perfectly well. Three times he returned .... and each time the recording seemed to have disappeared. He once worked for aweek with a German television team filming healers in the Philippines, only to have the lights fail during the crucial few-minute treatment of a patient who had been specially flown in for the programme. There is an almost willful elusiveness about phenomena under scrutiny that often leaves workers with lame excuses, and leads to a lot of ribald .... comments from their critics. Page 206. (This too is simply another example of the mischievous nature of the GW – Bob Dylan's 'Jokerman' – the motor at the centre of the phenomena, which refuses to co-operate with experimenters trying to neatly compartmentalize it.)

(d) In experiments................................, twenty dowsers failed to find buried ………..mines, and were unable to determine whether or not water was flowing through buried pipes. But it is difficult to ignore 'on the job' results by professionals going about their trade. The pharmaceutical company Hoffmann-La Roche ……. includes a dowsing survey when setting up new factories anywhere in the world. Their spokesman explains: "We use methods which are profitable whether they are scientifically explainable or not. The dowsing method pays off." Page 249

(e) I have had a forked twig twist in my hand with enough force to shred its bark ……. An American electronics engineer measured the movement of such a twig with a strain gauge bending beam and decided that the force was externally applied. (*A PK effect by the god within and BP – not the hands of the agent.* [My italics]). Page 249. From 'Beyond Supernature' by Lyall Watson. Hodder and Stoughton. 1986 ISBN 0-340-38824-2

12. Summary of material on 'Tribe' website. 'How to create a ghost'. Colin Brookes-Smith, a British engineer, together with Kenneth Batcheldor, a British psychologist, working on the assumption that poltergeist and similar phenomena were simply creations of the mind decided to do an experiment designed to produce effects similar to those seen in poltergeist manifestations. Under the guidance of Dr. A.R.G. Owen of the Toronto society for psychical research, they collected eight people and placed them round a table as in the traditional séance. They were then asked to invent a person who had lived around the year 1600, carefully describe his physical appearance, lifestyle and background, and then treat him as if he really had existed. They invented Philip, an aristocrat who lived in a castle set in spacious grounds. He was married to a cold, frigid woman who brought him little happiness and one day he was walking in the grounds and came upon a beautiful gypsy girl with

whom he fell in love. He enjoyed several months of bliss with her but his wife found out, and condemned her as a witch who had put a spell on her husband to make him love her. The lady was tried, found guilty and sentenced to be burned at the stake, but the man was too afraid of losing his castle and income if he protected her so he let it proceed. Some weeks later his body was discovered on the ground at the bottom of a tower from which he had leapt, full of remorse for his failure to prevent his lover's death.

The group tried for a year with the room lights on, to contact their invention but without success. So then they turned the lights down to simulate a traditional séance setting, put pictures of 'Philip's' castle on the walls and waited for results. This time they were lucky. After a few days one of the instigators of the experiment made a loud 'rap' ring out by covertly striking the table, thus 'priming' the situation and securing the group's belief in the reality of the phenomena. Soon after, believing that a paranormal effect had occurred, they asked 'Philip' questions about his life and circumstances which he answered by rapping the table, (this time it was a true paranormal affect – not the designer of the experiment) and gradually he assumed a definite personality, but his answers to their questions never included anything that the sitters themselves did not already know, such as historical facts. This lent additional weight to the supposition that psi was simply a product of the sitters' minds, and that anybody could do it. Soon after though, the rapping sounds which Philip made were supplemented by levitation of the table; this got quite extreme and it occasionally rushed over to greet latecomers, even pinning one member in a corner of the room, or dancing on one leg.

When they tried asking Philip to do things for them, he did so, dimming or increasing the lights on request and when asked to repeat a cool breeze which blew across the table provided it immediately. When asked to materialize, however, he couldn't or wouldn't do it. The object of this fascinating experiment was to find out whether paranormal effects were simply a product of the mind. The scientists who witnessed it were, however, amazed by the astonishing psychokinetic effects and were totally at a loss as to how they could be explained.

This is where the Theory of Material Mind takes over and explains the phenomena perfectly and it all hinges on the mischievous nature of the god within. At the moment of the first artificial 'priming' rap, the belief of the sitters would be secured, and they would happily co-operate to elicit the hoped for results. As the experiment proceeded the gods within- the motors of the sitters' unconscious minds - would be perfectly aware of the questions that were asked of Philip and, through the medium of bioplasma, would pool their resources to obligingly rap the table in answer to questions, such as Philip's favourite food or artist, whether he liked sport or was an intellectual, thus generating Philip's personality and the associated psychokinetic happenings. Essentially, they would be playing a game with the sitters and leading them up the garden path, knowing that there was no such person as Philip around at the time.

The original document gives a much more detailed account and I would recommend that it be read, since this will give a far more complete understanding of my argument. It can be found on the website 'Tribe', 'How to Create a Ghost'. Oddly enough, the original hypothesis, that psi phenomena were creations of the mind, would be correct, because the god within is an essential component of the personal unconscious, and the effects were obtained through the offices of the conscious and unconscious minds working as one. This, of course, would not invalidate the classification of the phenomena as paranormal

13. Extracts from 'The After Death Experience' by Ian Wilson.

Dr. Elisabeth Kubler-Ross, who along with Doctors Moody, Sabom, Ring, et al. has assembled her own collection of near-death experiences. Among these have been some totally blind individuals, of whom Dr. Kubler-Ross says:"We asked them to share with us what it was like when they had this near-death experience. If it was just a dream fulfilment those people would not be able to share with us the colour of the sweater we wear, the design of a tie, or minute details of shape, colours and designs of the people's clothing. We have questioned several totally blind people who were able to share with us their near-death and they were not only able to tell us who came into the room first, who worked on the resuscitation, but were able to give minute details of the attire and clothing of all the people present, something a totally blind person would never be able to do."From 'The After Death Experience' by Ian Wilson. Page130. Sidgwick and Jackson 1987. ISBN 0-283-99495-9

14. Extract from the writings of Steven Rose.
Indeed there is an entire, though often unregarded – nervous system in the gut – the enteric nervous system- with getting on for as many neurons as the brain itself (yes, we do think, or at least feel with our bowels,....). The 21st Century brain by Steven Rose. Page 64. Vintage 2006. ISBN 0-099-42977-2

15. The Astrologer's Handbook by Frances Sakoian & Louise S. Acker. Pages 49 – 52. HarperPerennial 1973. ISBN 0-06-272004-X

16. Extract from J.P. McEvoy.

1.

- (a) This (violation of Bell's inequality) means that in spite of the local appearances of phenomena, our world is actually supported by an invisible reality which is unmediated and allows communication faster than light, even instantaneously.
- (b) Interactions under non-local reality.

2. It can act instantaneously (faster than the speed of light).

3. It links up locations without crossing space.
- (c) The only popular examples of non-locality which immediately come to mind are the voodoo interaction of Haitian-African folklore, and perhaps extra-sensory perception.
- (d) Bell's work, which should apply to any fundamental theory of Nature (e.g. TMM - author)...could turn out to be one of the most important theoretical ideas of this century.
- (e) there now appear to be certain loopholes in experiments like Aspect's. These loopholes have reverted the proof of Bell's theorem to that of an open question. Einstein and the EPR still lives! (This 'non-locality' simply means that the mathematically

forecasted result of the experiment is not actually realised when the experiment is done. The actual value delivered is ½ where theoretically it should be 5/9.[author])

(Introducing Quantum Theory by J.P. McEvoy. Icon Books Ltd., 2007 Pages 170-171. ISBN 978-1840468-50-2)

17. Extract from 'the holographic universe' by Michael Talbot.
Another aspect of OBEs is the blurring of the division between past and future that sometimes occurrs during such experiences. For example, Osis and Mitchell discovered that when Dr. Alex Tanous, a well-known psychic and talented OB traveller from Maine, flew in and attempted to describe the test objects they placed on a table, he had a tendency to describe items that were placed there three days *later*.

The Holographic Universe by Michael Talbot. Page 237. HarperCollins 1996. ISBN 9 780586 091715

18. Extract from Cleve Backster.
Could the plant, he wondered, be sensitive on a cellular level all the way down to the death of individual cells in its environment?

On another occasion the typical graph (on a lie-detector or polygraph) appeared as Backster (a lie-detector expert) was preparing to eat a bowl of yoghurt. This puzzled him till he realized there was a chemical preservative in the jam he was mixing into the yoghurt that was terminating some of the live bacilli. Another puzzling pattern on the chart was finally explained when it was realised the plants were reacting to boiling water which was killing bacteria as it ran down the waste-pipe of the sink

Convinced he was on the track of a phenomenon of major importance to science, Backster was anxious to publish his findings in a scientific journal so that other scientists could check his results. But personal involvement in his experiments and even prior knowledge of the timing of an experiment was often enough to tip off a plant into non-co-operation.(this is obviously another case of mischief by the intelligence originating in the plant) He realised he would have to devise an experiment in which all human involvement was removed. The entire process would have to be automated. The test he finally chose, after two and a half years of trial and error, was to kill live cells by an automatic mechanism at a random time when no humans were in or near the office, and see if the plants reacted.

From The results of these experiments showed that the plants did react strongly and synchronously to the death of the shrimp in boiling water; and the automatic monitoring system, checked by visiting scientists, showed that plants reacted consistently to the death of the shrimp in a ratio that was five to one against the possibility of chance. The experiment and its results were written up in a scientific paper published in the winter of 1968 in volume X of *The International Journal of Parapsychology* under the title of 'Evidence of Primary Perception in Plant Life.

* * *

ADDENDA

1. Although, strictly speaking, it is not relevant to the theory, I would like to present a conclusive 'a priori' proof that there is truth in 'the art and science' of astrology. The proof came to me in 1975 and it hinges on the fact that the lives and actions of everyone on the planet interact and influence each other. The proof follows:
Let A influence B and B influence C, then A must influence C.
Now substitute 'the motions of the planets' for A, 'the actions of the believers' for B (which will be influenced by their belief in astrology), and 'the actions of the sceptics' for C.
Conclusion. The motions of the planets will influence the actions of the sceptics whether they like it or not. We are all, believers and sceptics alike, influenced by the motions of the planets and other celestial bodies. All life, human, animal and vegetal and even the population of the afterlife is influenced since it is part of the BPE which unites allliving things. This argument is true whether or not there are any direct physical or emotional effects of the planets on our lives and it may actually constitute the basic mechanism on which astrology is unknowingly founded.

2. In ancient Greece the body was thought of as having three regions, head, heart and gut. The head was associated with intellect, the heart with anger, fear, and pride and the gut with lust and greed. (fairly close to my own description of the functions).

3. I have not researched the subject but I believe that it is a fact that animals don't turn round, as humans do, when stared at from behind. This is because they don't have the 'sapiens' item as in 'homo sapiens'. They don't know that they are going to die and they don't feel the need to

cover their genitals.. Also, when communicating within the group, they don't look up to the sky to seek inspiration. This is simply an interesting reference to animal mentality, which could be a rewarding subject of research and impinge favourably on the study of human mentality.

4. An illustration of the nature of the material mind follows. A girl is waiting with her friend, Alice, for herboyfriend to collect her in his car. She is nervous and restless and confides to her friend, "I don't know what I'm doing, I'm all over the place without him". This simply means that she is feeling very fragmented; her mind, PP, is floating around in a distorted, random manner with nothing, no clear, fixed point or idea to attach itself to. When Jack turns up its various fragments and shapeless volumes come together and focus on him, joining with his mind, and she cries out how happy and relieved she is to see him.

5. It is a fact, I am now quite certain, that people do derive an inner warmth from each other, which is none other than the ubiquitous PP. I have seen many instances of this. Once, a few years ago, I was in my studio (my main occupation, for which I trained being visual art) feeling very cold and lonely. I had the feeling that someone else was there in another partitioned space, (it was a group studio where each artist had his own space). I called to him and he appeared about 20 feet away. Instantly recognizing him as a friend I felt a surge of warmth from him flood into my body. It was obvious that PP and/or BP flowed from him to me and I no longer felt cold and lonely. I have observed other things which lend cogency to this idea. Many times have I observed girls in freezing midwinter with shoulders, neck, midriff and legs exposed to the cold, dressed in light clothes more suited for the summer. I feel certain that they just didn't feel the cold because they belonged to a very warm, close-knit family or community which had large reserves of PP.

6. The god within is the fire in the belly that fuels our every action.

7. The Greeks wondered long ago what matter was, and Democritus invented atoms. Science has long since confirmed their existence but once the atom was discovered it was natural to wonder what they were made

of. So protons and electrons, made their appearance along with nuclei and energy levels, gluons and quarks, neutrinos etc.. We now have the option of seeing these entities as particles or wave functions. Continual analysis and division can't go on forever. When this point is reached I believe, or would suppose, that it becomes fruitless to talk of further division; instead a region of mystery is reached where a particle goes through two different holes at the same instant, and we get terms like 'strangeness', 'charm' and 'spooky action at a distance', although this latter is now accepted as true. It is here that the presence of intelligence or divinity can be supposed or presumed. There is no doubt, in my mind, that the god within, the divinity that resides in the gut of every person on the planet is a fact, real and actual, not imaginary or figurative. It is an independent entity and knows what it is doing. It is also a fact that there are other, discarnate divinities, some accompanied by spirits of departed, others perhaps waiting for incarnation.

8. The theory also explains why some days at the office or coalface are difficult and stressful while others are clear and free flowing. At the office, for instance, one is attached by dozens of fixed or fluctuating PP bonds to other workers in the office, and these bonds can be easy and relaxed while others are strained, stretched and twisted, thick, heavy, light or clear, and they all have their effects on the thinking and feeling of the room's occupants; (for an example of PP attachments see 'group minds' chapter 7 item 8) ; their entangled minds slow down and obstruct each other's physical and mental actions. Looked at from a less purely objective, non-scientific angle, one could express this as that when certain people were around one had difficulty concentrating. Getting on with workmates is sometimes hard work. I remember an incident in my early twenties when I found, in my place of work, that I couldn't move forwards or backwards. I was forced to stand impotently on one spot for about half a minute. This was long before my psi researches and I had no idea what caused it. I remember too,that many years ago I met a lady who told me that, as a patient in a mental hospital, her problem was a telepathic connection with Prince Rainier of Monaco which she could not shake off! It is also a fact that one may sometimes start feeling ill or 'queasy' and one's instinctive desire is to get some fresh air, so one goes

into the garden and immediately starts to feel better. This, I believe, is because the PP in the building, even in an empty room, may be very dense due to large numbers of people in the room or other parts of the building. It could be a fact that I myself am particularly sensitive to these circumstances because of the fact, and it is a fact, that I form telepathic bonds with unnatural strength and swiftness as a consequence of my accumulated knowledge of psi.

9. When I was working in an electronics test engineering department, age eighteen, I sometimes saw or felt what I used to call 'a holiday atmosphere'. Everyone was in a good mood and I remember that I could almost 'see' (with my mind I suspect), when walking down the alley towards my place of work, a warm pinkness in the air. This happened, I think, because the PP attachments, at the time, with my fellow workers were positive and mutually supportive.

10. Users of cannabis sometimes find that when a piston-engined aircraft passes overhead they have a sensation of having their mind being carried away. (One's mind, or part of it, I suspect, *is* being carried away in the form of a streamer of PP being dragged away from the main mass of one's mind). Cannabis must affect PP in some way.

11. I have just now discovered in my journal, an entry regarding evil spirits, which, at this juncture in my researches I see to be erroneous. I had concluded that they do exist and were very much to be feared. Now, however, looked at from an evolutionary viewpoint I ask myself can a baby, the first embodiment of any particular spirit, be born evil? An evil baby? The answer has to be no, although the nature of a child *can* be distorted in its formative years. So what motivates people like Hitler and Pol Pot? Political idealism is one candidate, or on a lower, interpersonal level, accidents and misunderstandings leading to false conclusions about motives, which leads to desire for revenge, feuds and warfare. Such things, including childhood influences, I think, are the origin of such things as bullying, torture, sadism, senseless cruelty, rape and pillage in warfare. Their qualities are exactly opposite to those of the GW which is pure goodness tempered with a certain necessary degree

of mischievousness. Thus the original pure spirit of the child is either retained or, in some cases at least, becomes adulterated with age.

12. In my lonely period as mentioned earlier, in addition to other things, I sometimes heard a voice saying, "thoughts are tiny feelings". Seen later in terms of knowledge gained (addenda 107) this implies that thought can exist only as a larger or smaller quantity of variegated feeling, and that at bottom, feeling is all we have, all there is; not counting, however, Plato's forms which are totally immaterial or metaphysical.

A feeling, therefore, can be one of two things, a percept or a blob of PP. The latter case is basic to the mechanics of gradual teleportation in which the teleported body is made up of millions of such blobs (I think!). In cases of instantaneous transference I am not sure what happens, but it is probably connected with the time/space transcending qualities of the GW.(See chap.5 item 13).and, possibly, the Platonic forms.

13.. My conception of the god within as gregarious, seeking always to join with others, is borne out by the fact that it is the origin of enthusiasm, and enthusiasm is known to be 'infectious'.

14. The BPE forms the basis of a group mind which stretches back to the dawn of life on Earth 3.5 billion years ago, encompassing and joining every speck of life that has ever existed on the planet. PP has been part of this mind since its beginning and together (BPE+PP) they form the 'psychosphere' (PS). It is by teleporting through this medium that travel into the past should be possible, but the time covered by this method can not exceed the age of the organism for reasons shown in the text.

15. The human organism acquires information via the senses, eyes, ears, nose etc. If it does not require a rapid response, as would be the case if danger were present, the information sinks to heart and chest where it is processed before being sent back to the head where it is recognized, evaluated and classified.

16. A fly lands cheekily on the tip of my nose. All life is continually struggling for survival and wellbeing. A fly lands on the tip of my nose

when I am feeling reduced, humbled or lacking in self-esteem. This has happened many times. The fly goes naturally with the flow of PP, and I always know, before it appears, that I am in 'losing mode'. I think that this happens because the PP density or, more likely, pressure, at this point on the body, is low at these times. PP flows, is exuded from the body and when the subject is well, must exit at and around the nose, as well as other parts of course, at greater volume and pressure than when he/she is unwell or low in mood. One is reminded of the concept of 'meridians' in acupuncture.

17. My powers, which I intend developing, strengthening and refining, are largely the result of the knowledge acquired in my researches as shown in this essay, and I am fairly certain that they can be taught, although a certain degree of personal skill and self-knowledge may be needed, and some people will naturally excel.

18. All gods within are connected and together they form the universal unconscious, which is an amalgamation of the collective unconscious (of the present life) and the collective unconscious of the departed in the afterlife.

19. Early in my lonely days one of the expressions that kept coming into my mind was, "all things are joined at the source", which I assumed meant that at my birth all my perceptions, William James's 'booming, buzzing confusion', must have been just that, one chaotic mass of information which had not had time to differentiate into separate parts. This expression I have taken notice of in developing my theory and it appears as 'the interconnectedness of the universe' in chapter 1. This relates very significantly to the fact that in quantum physics it is impossible to observe isolated subatomic phenomena without affecting them: the 'observer' no longer simply observes, he participates. This is a reference to the ubiquitous 'Heisenberg Uncertainty Principle' which states that it is impossible to know both the velocity and position of a moving particle at the same time, because any attempt to measure either of them will involve exposing it to a source of illumination which will impart energy to it thus affecting its value.

20. In anticipation of accusations of naivety in references to biology, psychology and other areas, I would like to make it clear that my only professional training is in the areas of fine art and telecommunications. I see biology as probably the most relevant subject for study in connection with research into psi, followed closely by physics, since psi is, of course, always associated with life and living things.

21. It is possible that some of the phenomena described are truer or more evident in my own case than with the ordinary 'man in the street', e.g. my ability to gain immediate telepathic contact with someone simply by thinking of them. These powers have been acquired through my own thinking and the strange places and situations that I've found myself in. I will try, in future, in addition to relating exactly what happens to me, to cast my mind back to a time before I had this knowledge, and give an imaginative account from this 'normal' standpoint, when possible.

22. I have designed a number of experiments which should provide an introductory test of the theory's validity. These may be refined and added to by interested parties, but will almost certainly be made unreliable due to the mischievous nature of the GW.

> (a). The vacuum wheel psychoplasma detector. (already described in main text)
> (b). A personal test of my own powers of establishing contact with any specified person. That is, the tester would choose a person known to him but unknown to me, a friend perhaps, give me a photograph of him, and I, by looking into the eyes on the photo would establish contact. The person contacted would become aware of some alien presence in his mind and, probably, as on other occasions when I've done this experiment, I would 'hear' in my heart, the words 'go away', at which point I would make a note of the time and date and then transfer my attention to something else. Later, the tester would confer with his friend to verify or disprove that a contact was made. This could be done with many tens of subjects, all on one day so it would constitute a

very rich and economically rewarding exercise. I think this expt. probably would not be subject to the unreplicability effect.

(c). An experiment to confirm the concept of UMSP (poltergeist phenomena) A statistical comparison of happy and unhappy pubescent or adolescent children and related UMSP in their households. (see item 4 Chapter 4). It would be necessary in this experiment to locate 100 or more families which display UMSP, and 1000 which should be chosen at random. One would then count the number of unhappy children in the UMSP families, and compare the result with the average to be expected in any group of 100 families. This would be calculated as follows; divide the 1000 families chosen at random into ten groups of 100 each, and count the number of unhappy children in each. Then add together all the numbers of unhappy children and express them as a percentage of 1000 thus obtaining the average. If the percentage of unhappy children in the UMSP families were significantly larger than the average, the analysis of the phenomenon would be supported. This experiment could be performed hundreds of times with different protagonists in several different psi laboratories, thus rendering it more accurate. This experiment, if based on already available figures, should not be subject to the 'non-replicability effect', although the GW could interfere with the inked figures in the way that it repeatedly interfered with the film in my camera in an earlier experiment.

(d). An experiment to confirm that PP exists and that members of a friendly community are less likely to feel the cold in winter than lonely people who live alone, thus confirming that a person who is deeply 'immersed' in a pool of communally shared PP is protected against the cold, whereas a lonely person is not. (see item 5. Addenda) The experiment should be performed on a cold day in winter and could take the form of a confidential questionnaire in which selected people are asked whether they were cold and lonely or part of a friendly communal household, and about their sensitivity to the cold. The questionnaire would need to be sensitively worded to secure the cooperation of lonely types, and prizes, in the form of a draw,could be offered to get as many responses as possible and to increase motivation which has been found to amplify positive results.

The questionnaires would be filled in at the end of the specified day, and each subject would describe as completely as possible, what clothes he/she had worn that day. If it were found that lonely people had worn more clothes than the others, my theory would be strongly supported. The experiment should be performed with as many people as possible, thereby increasing the accuracy of the results. This exp. Could be done in one or two industrial sized meat-refrigerators, which would be more convenient and more accurate though somewhat macabre and risky.

The above experiment could be repeated using only girls and these should be aged between sixteen and twenty-five. This would prevent misleading results which could occur due to age differences. Also my observation of people out and about on the streets suggest that dressing lightly in cold weather is much more common in girls and young women than in men.

(e) An experiment to confirm the existence of the afterlife. (already described in main text. Item 6 chapter 4)..

23. The concept of 'to feel is to know' (see glossary of terms), is inherent in telepathy (far feeling), and proven by the fact that telepathy itself has been thoroughly tested for and found to be factual though somewhat unpredictable. It is also illustrated by the phenomenon of making a distant person turn round by staring at his/her back.(See fig. 12).

24. Another illustration of the Primacy of Feeling. It seems to be a law of life that in any organization bureaucracy tends to encroach upon the freedom of the workplace thereby resulting in bad feeling and acrimony. Bureaucracy here, of course, represents logic, and the workforce the feeling and suffering organism, whose welfare officialdom should always keep in mind.

25. It is instructive and clarifying to think of the human being as having one body and two minds; The conscious one based in the heart with

a large store of memories, and animated by the spirit which works in tandem with the head; The unconscious one is based in the gut, also with a large store of memories, and motivated by the god within which works in collaboration with the spirit. It is clear that the head, which would asppear to be quite logical and independent, is not always so, since it works in tandem with the spirit, and, through PP the GW. It is also clear though, that in difficult and confusing circumstances, the head is the final arbiter.

26. Many years ago, in the middle of a mental illness, I often got the words 'deep underground' coming into my head. I see clearly now, that this was an intuitive indication that BP, as yet undiscovered, penetrated and existed many metres below the surface of the earth where a considerable amount must remain which belongs to no one mind in particular, but is part of the collective unconscious, symmetrical to the CC, available as an information source to any personal unconscious.

27. 07/04/09. Standing in Greenwich Park, one sunny day. I saw a very pretty little girl, about nineteen, walking down the hill towards me. I wondered whether I should speak to her but as she came level I decided not to. She passed behind me and as she did so I suddenly had a strong feeling of, to put it as nicely as possible, my private parts being 'jiggled' by a hidden hand. I realized instantly that it was the girl. When I dismissed her from my mind and turned away, she must have thought, "who cares anyway?" and laughingly, cheekily, in her mind, jiggled them, and I, turning round, saw her skipping off through a nearby gate. She would have been quite unaware that I felt it, but I did so because her imagined act, working through the materiality of her mind and imagination, would have disturbed the PP surrounding and permeating my private parts. That is, both she and I were immersed in a clear volume of PP which pervaded our bodies and would have enabled her thoughts and act of imagination, to connect with and affect my body.

Analysing this further it would seem that information was transmitted from her mind to my body, which raises the question, how long did it take for the information to get from her to me? I believe that BP and consequently PP, possess mass, and if this is the case they must also be subject to the effects of inertia. In another kind of system, say an auditory system, waves of pressure would originate at the source and would propagate through the medium at a speed determined by its mass and elasticity or viscosity etc. But I am sure that neither BP nor PP work by the generation of waves, or 'vibrations' which means, I think, that inertial effects do not apply in this particular example.

So what is the mechanism? I suspect that PP is full of information loosely carried or sometimes packed to a high density. The spirit is the identity, the essential self, of the individual, so the S component of PP is actually an extension of the person and wherever it exists there also the person exists, and she can make her presence felt. Thinking further, however, if the S were actually an extension of the girl herself the effect on my body would have been exactly simultaneous with her thought, which is quite possibly true, but due to the absence of suitable monitoring equipment, not capable of verification as yet.

It is a fact that in physics a period of time must elapse between transmission and reception of a signal, (ignoring Bell's inequality theorem), but we are not dealing here with physics alone, we are in a universe which admits the reality of metaphysical events, which allows the possibility that psi can travel faster than light. At this point I will drop the subject for a time and refer the reader to a more detailed treatment of Bell's theorem and the ' EPR paradox' in Chapter 5 item 13.

28. Whenever I, in my experiments, am called upon to perform similar actions, 'thought experiments', I know that the effects are often felt by the subject, so I have to be very careful. I always know, through feeling, when I've made contact, and sometimes I can avoid doing so by keeping a firm grip on my PP, i.e. my mind, to stop it making contact. (Sometimes the contact can be mildly distressing to the subject).

29. 'It's in the air'. This expression is used to indicate that the stuff of new theories, new music, new novels, new poetry, plays, inventions etc. new scientific and artistic movements, the 'spirit of the age', is ubiquitous and surrounds us. This is true not only figuratively, in the form of words, memories, conversations, plans, but literally 'in the air' in the form of

the CC and PP which, I believe, is packed with information available to anyone.

30. An example of the Primacy of Feeling. Writing a poem one day, (one of my favourite activities), a jumble of words came to me which had no literal meaning, plot or structure but simply *felt* good, full of rich feeling, hidden ideas and resonances. I finished the poem, including these words, and put it aside. Two years later I found it by accident in some papers having forgotten all about it. To my surprise I saw a rational explanation for this 'blob of feeling'; all the events and contingencies that went into its forming but were ungraspable or beyond analysis when I wrote it were expressed by the psyche as a satisfying blob of pleasurable feeling; a meaningful whole. That is how life felt at that time but now I can see the reasons for the feeling; the events and connections, the *plot,* that went to make up my life as it was in those days. This is an excellent example of 'the primacy of feeling' operating to preserve truth which would otherwise have been lost. There was no rational meaning to the words, I knew not what they meant, but they *felt* good so I kept them. The rational content was perceived later.

31. The hippy expression 'get it together' is an intuitive perception that in this situation one is gathering one's PP powers together to focus on something.

32. Standing at the bus stop, freezing cold day. A girl standing two metres to my left, back towards me, no eye contact. After about five minutes I suddenly became aware that it felt as if a region of warmth had started to form about halfway between us. It got stronger and stronger and it became obvious that it was a PP bond developing between us. Her bus came. She left. It was an example of a bond growing as a result of prolonged physical contiguity and consequent conscious connection, starting with unconscious connection between gregarious gods within, which would already be connected through BP in the local unconscious.

33. The GW is at the seat of feeling, the stomach and gut, where BP is densest, but its effects are also felt at heart and head levels where BP is

a component of PP. At heart level it is responsible for strong, sometimes emotional feelings, and at head level it registers finer or more delicate, nuances of feeling which one could call 'mental feelings'. True emotional feeling, i.e. anger, sadness, hope, happiness, despair, relief happens in the middle of the body at heart level, and involves the god within and the spirit.

34. Tough, fighting, aggressive people have very thick, strong, PP with which one can become easily entangled, whether one likes it or not. Intellectual types have gentler, more diffuse, clear, tenuous and attractive PP.

35. It is possible that PP has a perceptible though very tiny effect on the aerodynamics of a mechanism.

36. Experiments in particle physics take place in a clear, tenuous volume of PP and at this point the GW and the unconscious must interact with matter (particles) via BP. Perhaps some of the particles observed are particles actually belonging to the BPE; if the BPE could be shown to be particulate this would indicate a very significant overlapping of the physical and the metaphysical, although, in view of the ubiquity of the god within, a non –particulate phenomenon, I think, they would overlap anyway. It is in any case, interesting that the experimental results provided by two Russian physicists strongly suggest the particulate nature of the BPE. (see discoveries of Grischenko and Inyushin Chapter 1 item 6 Composition of bioplasma).(Ref. 7)

37. This entry duplicates part of entry 41 so I have decided to leave it out rather than embark upon the lengthy procedure of renumbering.

38. The popular conception of telepathy is that it is the transmission of words or images, which, of course it sometimes is. The general truth however is that it is the perception of feelings, (far feeling) awareness of mental connection with another person(s), sometimes of how that person is feeling. This can lead to the transmission of discrete information (feeling is continuous). That is, ideas, attitudes, thoughts can be inferred from the feeling, ('To feel is to know'. Chap. 3 item 6) but thinking

back to 'PP recordings' (chapter 1 item 6) it would seem probable that PP might contain discrete information in some form.

39. In my period of extreme loneliness a thought sometimes occurred, as mentioned earlier, which amongst other things stated, as I held a roll of my slightly overweight abdomen in my hand, "this stuff thinks". This was, I think, a case of me – my spirit – making an observation, and it suggested that thought originated in the cells of the body whose other function was the secretion of bioplasma. I think it would be correct to think of 'cell thought' as different from, less conscious, but more 'macro', than 'conscious mind thought', which, I believe, is more 'micro', and originates mainly in the heart, as does almost all other voluntary mental phenomena. That is, the unconscious mind, I believe, is far bigger than the conscious one, and it is always joined to other near or distant ones.

40. It could be claimed that the inclusion of divinities in my analysis is simply a 'cop-out' (if you can't explain it just say it's magic) but this is far from the truth. My first encounter with the idea of the god within was through the writings of Arthur Koestler, and as I progressed through my life, I gradually became aware that it was real, existing and active in my gut, not just an imaginary or poetic concept. Everyone has a divinity or intelligence in his/her gut.

41. Freud's discovery of the unconscious was to reveal the independent working of a real, objective entity and was his most important contribution to science. The id, ego and superego were invented to structure the manifold variety of psychological phenomena. They have no objective reality, except in the sense of their parallels, i.e.(a) basic lust or desire, (b) the conscious 'I', and (c) the conscience, which could be seen as:

(a) the God (or intelligence) within
(b) the essential self or spirit + the head
(c) the GW + the spirit + the head

In this capacity of the conscience all three components must agree, I would think. The theory of material mind also deals with objective entities. The god within is the motive force of Freud's unconscious revealed, and it is real, objective, and independent as is the spirit in the heart.

42. There are basically two ways in which travel into the past is possible:

   (a) Temporal Teleportation from position A in the PL to position B in the past lifetime of the traveler.
   (b) Spiritual migration from position A in the AL to any other place in space or time except the future.

Time travel into the past in case (a) is possible because PP extends unbroken to the dawn of life on Earth, and time travel in this scenario involves teleporting from the present, through the BPE, into the past, growing younger as one does so (diag. 4). I will explain. Let us consider me, sitting here at the computer, at 22:00 pm on Monday 01/02/2010. I am surrounded and permeated by a plethora of PP which fills the room and connects me with others outside my room, (since walls etc. are transparent to PP). I am directly aware only of myself, the walls of the room and things here in the room. This plethora of PP will be subject to various forces i.e people moving about next door, or fluctuating telepathic connections, but it will still be there - there is no point in time when there is no PP present - at 7.00 pm tomorrow. In other words, there will always be an unbroken connection, day to day, and year to year, right back through the days and nights, right through prehistory and the various geological ages, to the Cambrian Age and the dawn of life. At such distant times the BPE will be extremely tenuous since the only life forms will be bacteria but by the Devonian Age, 'the age of fishes' it would be getting denser to arrive at a plateau where PP density equals at least 0.001 which should remain fairly constant right up to the present day.

I don't think it would be possible to go back further than one's own youth in the PL or perhaps early childhood, because in this kind of time travel one is essentially travelling back through one's own years of growth, and would ultimately lose awareness of oneself as a conscious being if it were attempted. If one were to actually do a trip by this 'teleportation' method, into one's own past, say to age fourteen, one would arrive at age fourteen and everything, animal, vegetable or mineral would be exactly as it used to be at that point in time. Moreover one would not be able to return immediately to the present; one would have to grow all through the intervening years, live one's life over again, to arrive at one's present age. (This idea appears to nullify freewill since one would have to make the same actions and choices to arrive back here at this precise spot in 2010). Pushing this supposition to its limit I could say that perhaps I **have** flipped back through time and am now growing back to age ninety-one or ninety-nine! I don't think, however that it would be like this – my return to the present would be more like a rivulet of water trickling through mud, taking the already established channel to the same destination i.e. the time at which the 'flip' was initiated. (See diag. 4).

It will be seen that this kind of time travel is quite different from that posited by physics, in which, presumably, one could go back to the time of the romans, retaining one's present age and appearance and appear leaning on the colosseum dressed in topper and tails. This latter scenario, however, exhibits a logical contradiction which the theory of material mind does not. That is, according to the physics method I could go back in time to shoot my grandfather etc., the consequences of which are obvious. Retro-time travel as occurring in the TMM time-flip involves no contradiction, but merely growing younger physically and mentally as one reaches the target time, where would be found one's grandfather and every body else, all 60 years younger. Essentially, the universe and everything in it, would be sixty years younger; that is one would have changed one's position in time. The weird thing here is that a single act committed by one self can result in NO!NO! I was going to write that this act results in making the universe 60 years younger – but that is wrong! It means that one sees, as I somehow did, (temporal clairevoyance?), a younger self, existing at a time when the universe and every thing in it **was** 60 years younger, and could if I'd been a little braver, or reckless, have done the flip. I feel quite sure that the departed in the AL, at least, **are** able to travel back to times before their birth because (a) the GW will take them anywhere they want and (b) I did a thought experiment which involved me mentally projecting a large part of myself into the AL, contacting it and some of its residents, so that I could travel back through time, but having passed beyond what I felt to be the bronze age I began to feel a cold loneliness creeping into me which made it it imperative to go back home to my own time and friends in the AL and the PL. Thinking back I see that I could very easily have got spiritually and materially involved with the people of those times, as I did with Beethoven as described a little later, and the Greek philosophers, with possibly disastrous results.

I have seen that here in the New Age practically anything is possible. We have almost unlimited freedom. The question here is 'how to do it? For instance, How does one change one's position in time?' My own experience was quite spontaneous and I must confess that I have no idea as yet, and can only say, keep an open mind, keep trying, and one day if you are lucky and have the patience, it may, as a manifestation of the New Age, happen to you. I know that the AL and PL can interact, not only through the activities of mediums, but through my experience of being attacked by Plato when I disagreed with him and through the other interactions with Heraclitus and Pythagoras. And it is not all one-way; I once found myself in a very difficult relationship with Beethoven, my first hero in my youth. Some how I had become physically connected with the flesh surrounding one of his eyes and it was only by taking great care that I managed to let the connection, which was a PP/fleshly one gently disappear without causing him discomfort. From this I see that material and spiritual connection and influence between the AL and PL is at all times possible, by virtue of the fact that the mind is material, and this connection is a two-way phenomenon; moreover this is possible because there are two kinds of stuff in the universe, physical and metaphysical. This strongly suggests that such time travel could be a very risky business, as observed, since one could get drawn from one's safe observational place 'in spirit' in the AL to an actual material interaction with the people or organisms one was observing, deep in the past. I say this in answer to the philosopher Julian Baggini who claims, like many others, that there is only one kind of substance in the universe. How wrong can one be! A related instance of this is the time I almost resurrected my parents by accident – I could almost see their faces!

This, retro-time travel in the AL, it would seem, is actually possible though it may entail practise and repetition to acquire the ability to initiate and sustain an expedition, and, of course one would have to die first! Travel into the past in the present life would necessarily be, as explained, limited to a few generations. The future is a different story. Having, just now, tried a thought experiment, trying to travel into the future, I felt the dry grip of old age crushing my heart, and hurriedly drew back to the present. We are of course, already travelling into the future and I, in my position between the PL and the AL, when I try to travel in that direction, tend to multiply my rate of getting old by an intolerable factor. I think I must conclude that travel into the future from a position in the PL is not possible, we are already going fast enough in that direction which ends only in death. I also suspect that travel into the future from a base in the AL is impossible. The future is largely a closed book I think.

This negative, however, would not preclude precognition, in which the god within, roaming beyond the laws of space and time could, at times, spot potential events useful or harmful to the organism. Accepting this one would not need to belong with the departed, one would be in the PL at the 'wavefront of advancing time', in other words*, **now.** The GW, however, would take one to any desired destination past or present, except the future, and anywhere on Earth, and ultimately, I'm certain, it will be our road to the stars and their 'exoplanets', in a non-local universe.(See chap.5 no.13)

If one considers the fact that material or psychoplasmatic intercourse between the current PL and AL and the past PL and AL is possible one might think that it is also possible for the future PL and AL and this would indeed be true – when the future PL and AL arrive. But, of course, we have already seen that travel into the future will probably remain impossible and the future will remain inaccessible except for the occasional instance of precognition. e.g. known future events – Blake's teacher hanged – Nostradamus's forecasts – premonitions – Dunnes 'An Experiment with Time'. The future Afterlife will exist, of course, in the future just as the past Afterlife and the current one do now and in the past. I do believe, as written, that from its position in the AL a given personality would be able to travel to any other place or time, except the future, e.g. the Greek philosophers with whom I made contact and the hominid from 7,000,000. years ago. On the face of it this could mean that a person in an OBE or NDE could also perform this exercise but the occupant of the AL would be there in his/her metaphysical totality, that is GW /spirit/mind, and free to roam at will, whereas in the OBE or NDE the GW would have to stay in the body, so that the spirit would be obliged to stay 'close to home'. It is interesting that this whole business of time travel through PP is similar to that of staying young the only difference being that in one case the action is in the PL and limited by time of birth and the other in the AL and virtually unlimited. It is even possible that the AL venture could take place where a lot of friendly spirits, archaeologists perhaps, grouped together for company, thus enabling them to travel beyond the bronze age without getting cold and lonely, i.e. PP getting too thin, or too far from home, into the various geological ages to see the dinosaurs, although it may be a century or two before that can be done or, nearer to home, the building of the pyramids. I do, in fact, suspect that travel into the past, in the AL would almost certainly be limited, at first, to a few hundreds or, at most, thousands of years even for a band of 'temporal explorers' since it would involve staying in telepathic contact with spirits and gods in the AL at 'home base' here in 21st century Europe and taking care not to get lost. The outposts marking the limits of exploration would be recorded or remembered in some way, so that the next team of explorers could push them even further. This would constitute an exact paradigm of geographical exploration in the present life here on Earth which remains incomplete to this day e.g. the Amazon and Borneo.

    Historical exploration of the past, on the plane of the PL, (defunct civilisations), is, of course, well established but since belief in the AL is far from universal, the proposition that it could be used to observe and explore past events, civilisations, real things happening while they happen, e.g. the building of the pyramids or stonehenge might only invite laughter and ridicule, despite the fact that it is literally true.

    This, despite its outrageous claims upon the reader's capacity to believe, is not waffle but the exact truth as I see it. There is one minor objection here as to travel into the future – it is probable that the future is simply a mass of unresolved possibilities, whereas the past has all happened and is capable of being explored. In the case of the future the

GW would probably refuse to oblige or to go very far, for reasons known only to itself, or unavoidable and unacceptable effects on the traveller.

However, as a last illustration of what will be possible in the future, assuming that it will not be possible to travel directly into it oneself, is that it should at least be possible to leave instructions with one's great great grand children perhaps, to perform a resurrection of one's body at a date far in the future, which would itself amount to an essay in time travel into that mysterious, unknown and fascinating place.

Now at this point I must provide solid material to support the somewhat fantastic claims made in the last few paragraphs. The main point to be made clear is that there is a two-way interaction between the AL and the PL and that where the AL is almost totally metaphysical the PL is material except for the fact that it is animated by the metaphysical entities of GW and Spirit. There will, also, always be, somewhere in the world, numerous telepathic connections involving genuine mediumistic contacts between the AL and the PL which constitute a further blurring of the edges between AL and PL.

Now I must go back to a period of very low self esteem in the care home where I used to live, very close timewise to my 'standing compulsion' recently described. At the time they were showing George Best's funeral on TV and somehow I became entangled with it and with George, who was in the AL of course, as I quite often do with others. I had a strong feeling that I should stop eating lunch since I'd had enough and eating more could be seen as gluttony; nevertherless I continued to eat much to my own disgust. I don't know exactly how this happened but I do know that I became spiritually and materially entangled with George, which I think displeased him, as I watched his funeral on TV, so much so that I began to fear for my own life – this took the form of a sensation in which, if I remember correctly, something was hovering above me and threatening to drag me into the afterlife almost as a kind of punishment for interfering in George's funeral. I should make it clear that the threat did not come from George but rather the AL itself. It became so strong a feeling that I was very hesitant about going to bed, fearing that I might die in the night. The essential characteristic of my life at this time was that I seemed to be rooted in the PL but was also partly in the AL and the whole thing was **real**, i.e. fairly dense PP, or even a mixture of PP and matter just as my interaction with Beethoven was **real, that** is a fleshly one. I could have shaken Beethoven's hand, in fact, believe it or not, we hugged each other! The essential point about all this is that it *felt real,* **in other words** *was* **real,** semi-material, and I can still remember it now as such. .

43. Although it is theoretically possible to perform a resurrection, it would necessarily have to be one of the resurrected one as he/she would have been after death; (My principle of the individuality of the spirit would make it impossible to have a resurrected person here now on earth while leaving an older version of him in the afterlife. If this were not so it would be possible to resurrect one's grandfather as he was at age 14 and see him standing there in front of one's desk. Clearly impossible). It would feel quite crazy and violate the POF.

However, if one did perform a resurrection, of one's grandfather perhaps, would he appear complete with the illness that killed him? Not necessarily, since, assuming that resurrection had long since become a common -place practice, as I believe it will, he would probably be brought, as an unconscious (GW) and a conscious mind, (head + heart), from a secure and comfortable position in the AL, to a position in the PL where he would materialise, by appropriating the necessary PP from the environment, thereby building himself a new body (as I almost did with my parents). This new body would be more or less identical with his body as it was when he died, but presumably it would be in good health. There is no doubt, of course, that it would be chronologically older than it was at death, although this would not necessarily show. The same reasoning would apply to resurrecting Mozart or Newton.

44. To continue with this analysis; assuming that a resurrection was performed, the resurrectee's friends in the AL would realize one day that he'd disappeared but after a few years (or whatever passes for time in the AL) he would be only a distant memory.

However, at resurrection he would appear, as written, looking much as he did when he died except that a long, presumably pleasant spell in the AL would have restored his health and well-being which would then be a mental thing since that is all one is in the AL – a disembodied mind, or rather minds, conscious and unconscious, the body does not go to the AL – but after years in the afterlife he would have forgotten his previous morbidity, so that his now healthy mind(s) would materialise as a a healthy body/person.

Depending on these factors he could take up where he left off and continue to live out a further stretch of life. When he died a second time he would reappear in the AL at the same age as when he first died plus the new lease of life that resurrection had afforded him. It would seem that eventually the faculty of backwards and forwards time travel and teleports to and from the afterlife may be as normal and commonplace as taking a walk in the park is to us at the present time. This, despite its mind-boggling aspect, is the truth as I see it. We are entering an age of miracles, which promises to be as advanced and different in its technology, from that of today, as the latter's is from that of Neanderthal man.

45. It occurred to me recently that time needs consciousness in order to exist. Interesting thought. Postpone.

46. A memory may come to mind from one's own or somebody else's unconscious or the collective unconscious.

47. I believe, in line with orthodox science that the concept of a creator God is fallacious. However, I do believe in the spirit and the god within, both metaphysical entities without which life, as we know it, would be impossible. I believe that we are alone in the universe, in the sense of being independent and thinking for ourselves, without guidance from higher beings, although there must be advanced life in other parts of it. Expressed in another way and quoting William Blake I believe that "everything that lives is holy" including ourselves, and even alien life wherever it is. We are conscious and enjoy a two-way relationship with the gods.

48. One of the chief differences between the theory of material mind and the theory of relativity is that the former is intrinsically bound up

with living systems and the workings of different parts of the human organism..

49. In view of the fact (which is what I believe it to be) that the god within will permit measurable, extended and replicable paranormal activity only when such activity is not being used solely to investigate its origin or analyse its behaviour, would it not be a good idea to construct experiments that perform some activity useful to society? A somewhat fanciful example: Use people who regularly and frequently experience clear OBEs to journey into foreign countries, e.g. Iran to monitor nuclear activities and report signs of bomb construction. (See chapter 4 item 10).

50. An extract from my journal 26/04/2004 relating to the gradual discovery of the material nature of mind: More evidence for bioplasma. In the pub talking to a friend; I held on to my mind (a mass of bioplasma?) but lost my grip as we talked. My mind seemed to be a cool, full, heavy feeling in the air between shoulders and floor level. As we talked, standing by the bar, I felt my mind go behind my friend, we were facing each other, to go creepily up his back. He exclaimed, "What the f***?" and span round to see what it was. He saw nothing of course, for this mindstuff was invisible. I must then have regained control of my mind and brought it round to myself. I remember too, way back in '78,'79,'80 walking down the road, alone in my loneliness, feeling that my spirit was modulating and playing with the – not yet recognised as such - bioplasma. It was a bit like bouncing a tennis ball. The air/space would modulate and bounce. This cool, heavy plasma is how, I think, the mind must influence objects through PK.

I remember also, many years ago when I was living with a certain crazy lady, that she sometimes complained that she had 'nothing to concentrate on', and couldn't get 'above herself'. I know exactly what she meant, for I had a similar problem. As she moved around the streets she had the feeling, if her mind could be symbolized or represented by a large, near-spherical water-filled, transparent polythene sack about two metres in diameter attached to her back at shoulder-level, that it was dragging along at ground-level. Normally, if one were happy and in good mental health, this 'sack' would float above one's head, full of helium perhaps, – it would be, as I see now, the conscious mind of the subject, finite as described, and it would float in the surrounding collective conscious lifting one into a state of normality and – hopefully – happiness, instead of dragging one down to the ground.

51. I have seen it asked 'how is it that people who claim to be able to contact the afterlife, always report contact with long-gone famous people rather than Joe Bloggs the plumber from 18th century Grub St. This can be compared with looking at a range of distant mountains, where the peaks represent Beethoven, Napoleon et alia and the invisible cottages in the foothills are Joe Bloggs and his friends. Why, indeed how, would I make contact with a member of the unknown masses when I'm not even aware of his/her existence?

This is similar in some ways to the fact that ghosts are always seen wearing clothes rather than in their birthday suits, and usually too, in clothes appropriate to the period in which they died. I have not yet found an explanation of this, largely, it must be confessed, because I haven't really tried as yet.

I am writing now several months after writing the foregoing, and a very cogent explanation of this phenomenon is that what we see when seeing a ghost, is not a physical body but a mind, the mind of the ghost, and it would appear as it imagined or conceived itself to be – not naked but wearing the clothes to which it was accustomed. Indeed it would have no conception of modern clothing for obvious reasons! It would know the clothes of its own historical period.

52. It occurred to me today that WW3, if nuclear, would not only eliminate all life on the planet – it would probably do the same for the afterlife! This would be a consequence of the AL being framed as in my second conception – bioplasma would be destroyed, I think. ***All*** bioplasma! Since BP is a material thing. However, the metaphysical components, spirit and god within – most certainly –would survive. Thinking on I saw that one could only hope that WW3 did not occur for – deadly thought - the soldier who went to the hopefully comfortable AL might find it to be very far from comfortable. If, indeed, the AL were destroyed by a nuclear war, one presumes that the PP would be

stripped from the spirits, which is like saying that the individual's PP was destroyed along with his/her mind, since the PP, in cahoots with the spirit, actually *is* the developed mind, or so I believe. This would still apply if the buddhists were right since the Great Spirit too would also be destroyed. However I think the buddhists are wrong on this occasion for the reasons given, namely, the philosophers from 5th and 6th century Greece are still at large in the AL as is the 7,000,000 years old hominid contacted by me. This whole question would still apply if the nature of bioplasma is as described by Inyushin and Grishenko (Ref. 7). One can only hope that WW3 will never occur because whatever its effects on the afterlife if any, it would amount to a terminal event for Earth. Whatever the truth may be this question still applies in the case of the self-immolating buddhist monks; their PP too would presumably be destroyed but they would have access to the rest of the PP remaining on Earth to build a new mind in the AL.

This question raises a whole heap of speculation as to the fate of PP at the scene of a forest fire or similar occurance, in fact it is becoming too speculative to spend any more time on right now, but must be thought out or experimentally settled later.

53. I am pretty sure that my second conception of the nature of the AL is correct, but it does suggest a host of consequential questions i.e. 'from where have the four or five billion new spirits born this century made their appearance?' Is there an infinitude of discarnate infant spirits awaiting incarnation? Etc.. Plato believed that the spirit existed before birth and became trapped in the body. I would suggest that the spirit is created at the same time as the body, i.e. spirit and body are born together as a unit.

54. Making music is an essentially social activity and 'infectious enthusiasm' is an essential component. This is recognised in the Chinese 'Book of Changes', the I Ching, where Hexagram 16 is enthusiasm, the typifying quality of the god within.

55. Raudive voices. I have long been aware of these phenomena but, like many other parapsychologists, I felt that they were akin to 'Leonardo's paint stain' or seeing faces in the structure of clouds – a purely subjective phenomenon.

Recently, however, I read, in finer detail, that Konstantin Raudive, a Latvian psychologist, has discovered that by speaking directly into a microphone and recording the output, recording the 'white noise' output from a radio tuned 'off-station' or recording the output of a crystal set with a short aerial, he can find a collection of softly speaking voices on playing back the tape. They speak in up to seven languages, all of them familiar to Raudive himself. This is not an accusation of fraudulence, but merely an interesting fact which should throw light on the origin or cause of the voices, which Raudive believes to be messages from the afterlife (which I, personally, know to exist but cannot prove as yet). The voices sometimes respond to questions posed via the microphone almost achieving, thereby, the status of conversations.

On 24 March 1971 he submitted to an examination involving an eighteen minute long recording of the effect in the studios of a major English recording company, all of it under the close supervision of engineers who checked for freak pickups from other radio stations and any other interference. A separate recording, in addition to the experimental one, was made which recorded every sound made in the studio. Raudive himself, although one assumes he was present, was not allowed to handle any equipment. When the tapes were played back the experimental one was found to have more than two hundred soft voices some of them loud enough to be heard by everyone present.

Telepathy between the current and the afterlife is a common occurrence and usually involves the services of a medium, although I myself have my own sometimes spontaneous, sometimes intentional experiences just as, presumably, Raudive did. I think what was happening in his case was that he was, actually, in touch with the afterlife and in some way actuated the recorder or modulated the tape by means of involuntary, unconscious psychokinesis. It is very significant, I think, that one of his contacts was his mother with whom his unconscious telepathic link would be very strong. I am indebted to Lyall Watson and his book 'Supernature' for most of the information on Raudive.

56. A disproof of Karl Popper's principle of falsifiability. My own a priori proof that there is truth in astrology, demonstrates that Popper's principle is unreliable; i.e. astrology is not falsifiable, therefore according to Popper, it does not qualify as a science. My proof, however, shows that the 'art and science of astrology', though not an exact science, and necessarily involving interpretive faculties in the practitioner, (as does the making of a diagnosis by a GP) certainly does contain a large measure of truth and cannot be dismissed as a pseudoscience by any open-minded examiner. I myself, along with Kepler, have a long-term dream of converting traditional or current astrology into a more exact and mathematical science based on probabilities.

57. The psychological state of 'fragmentation' is an uncomfortable condition in which the psychoplasmatic conscious mind is split or torn into several pieces, united by an extremely scarce background PP, usually resulting in a feeling of extreme fragility.

58. The TMM could be of great use in the treatment of mental illness, insofar as it could be used to analyze and clarify its nature and indicate curative procedures. Telepathy and UMSP, of which I could not speak in my own sojourn in hospital for fear of being accused of suffering from delusions, played a very powerful part in my own, often unjustified, detention. I also met a very intelligent undergraduate lady, who was herself studying psychology, who told me that her own trouble was centred round a disturbing and immovable telepathic attachment to Prince Rainier of Monaco! It is, actually, quite likely I think, that mental illness could be carried over from the present life to the afterlife, since its causes are often difficulties in relating to others, and one's new place in the AL might not change this.

59. In one's youth one's actions are much more conditioned by feeling than reason annd intellect, as are one's communications with others. That is, the weight or import of any verbal exchange is much smaller in comparison with the associated feeling and body language than it is when one is older, in one's thirties and forties, etc., partly of course, because one's 'knowledge possessed : feeling' ratio becomes much greater

with advancing years, and one will tend to rely on knowledge rather than feeling to guide one's actions. Here I think a quotation from Bob Dylan is in order. It goes:

> "While one who sings with his tongue on fire
> Gargles in the rat race choir
> Bent out of shape from society's pliers
> Cares not to come up any higher
> But rather get you down in the hole
> That he's in."

He is singing here of the way a 'loser' can manipulate, via PP, the feelings and actions of a motivated, aspiring person, to bring him down to his own level.

60. It has just now occurred to me that the afterlife must contain a massive preponderance of the old and elderly, unless, due to the 'age continuum effect' – the mechanism that allowed me to contact Leonardo as both an older man with flowing white beard, and as a young and friendly one, it is possible for a departed soul to choose to exist at anyone of the moments between its birth and death. This suggests that the apparent, seeming or felt age of any given member of the afterlife may be a function of the combined ages of the people – in the AL and the present one - with whom he is currently connected. Or, perhaps a dweller in the AL is able to choose the age at which he/she prefers to exist, or even flit between different ages.

61. Blank. (Was a duplication)

62 . The metaphysics of the Theory of Material Mind is, like the data of physics, and as is the theory itself, surprisingly, solidly founded on empirical data i.e. the OBE, PK, precognition, telepathy, remote viewing etc.. All of which are well-attested commonly occurring phenomena. The only limiting factors being (a) the data's unpredictability, and (b) the absence so far of a unifying theory, both of which this essay is putting right.

63. As a test of the theory's credibility it might be interesting to compare the origin of the universe according to the TMM (permanent, undying mystery of 'the gods'- gods do exist – this is an established fact which I have yet to prove), with that of the 'Big Bang' (nothing existed and then suddenly the universe exploded into being). Which seems more likely, the gods were there or nothing was there? (Is it possible that the gods created the big bang!?).

64. John Lennon's line "love is real, real is love" means exactly that. Many times have I wondered exactly what he meant by these words. I now see that he was referring to the materiality of psychoplasma. (Feeling is real as in TFITK. It is impossible to have a feeling, e.g. love, without being aware of it unless it is masked by other, stronger ones.).

65. At some other place in the essay I have wondered whether PP could have a slight effect on the aerodynamics of a system, but I have since seen that in practice it wouldn't. The reason for this is that when the designer did his measurements of airflow, he would get a figure which included the effects of PP if any, and he would be quite unaware of this as would his system, and would assume that his figures were for air alone, and in any case he would be designing within the presence of the basal density of PP which would probably be considerably less than the corresponding air density. Ultimately I don't know the answer to this one.

66. What is joy? What is love? What is hatred? What is a feeling? They are all '*felt*'. And ultimately they can be described and understood only in terms of the TMM.

Essentially, orthodox science hasn't the faintest idea of what they are and if asked would, at best, point to an electroencephalograph, a varying quantity on a moving graph of brain activity, and say that they were 'associated' with differing kinds of brain activity'. Here then, when hearing the magnificent 'non shall sleep' from Puccini's opera 'Turandot', the medical worker would point out the EEG's activity which accompanied it, and which would result in a zig-zag one inch 'peak to peak' tracing of the pen across the paper. He would also point out the physical accompaniments in the subject of increased respiration, heart rate, skin conductivity and perspiration etc.. This is the closest orthodox science can get to giving a full description of the constituents of a feeling, even though it has taken the analysis of matter to the extremes and subtleties that exist today.

Feelings can be divided into three kinds, (a) mental
(b) emotional
(c) Physical

An example of (a) could be Bertrand Russell's 'moment of intoxicating delight'at the successful solution of a mathematical problem. (b) could be a feeling of anger, joy or pride in which the proud one feels a swelling of the intercostal region. The last one, class (c) would be the physical pain of a burn or other injury to the body. In this essay I am more concerned with the former (a) and (b) than the latter (c) which is a very simple case of 'stimulus, electrochemical message to the brain, neurological brain activity, reflex electrochemical message to the relevant muscle or body part to take appropriate action'. The former are far more interesting and complicated involving, essentially, the relationship beween the god within and the conscious mind., which are the two nonphysical components of the organism.

I think, therefore, that any feeling (mental) can be shown to consist of (x) a constant striving on the part of the GW, to fulfil its duties and provide warmth and energy, in opposition to, or in accordance with (y) the spirit striving for comfort, security, positivity and control, amounting in the end to two minds in the same body struggling for peaceful coexistence. I'll return to this, hopefully, in my next essay.

67. In my lonely, alienated period mentioned early in the essay, as described, I used sometimes to get verbal messages or observations from I know not where precisely. One of them was, much repeated, "the inner truth is greater".I used to struggle to understand this, knowing that the scientific method required externally or objectively present data, empirically acquired, to produce reliable results.

It is only since completing the section 'the elixir of life', (Chap. 5 item 7) that I have finally seen what it meant. It meant that the attitude of mind is crucial in any human undertaking. Relating this to ordinary life it means that no matter how grim one's circumstances, one should not 'lose heart' but should believe despite the negative evidence, that happiness can be achieved again.

It is, in essence, the old religious command 'only believe, and the Lord will deliver thee' (from current penurious circumstances) but, of course, since belief in a creator god is now mostly absent, among scientists at least, its meaning, updated, is 'think positive, where there's a will there's a way'. 'The inner truth is greater' means that the mind is stronger than the body (Chap.2 item 5) and with true belief and application practically anything is possible. The principle is also expressed by the old adage, 'faith can move mountains'.

68. It seems to have escaped the attention of 21st century physicists, that Newton's Theory of Universal Gravitation, enthusiastically embraced by scientists from the mid 17th century through to the end of the 19th century, although essentially describing a large piece of clockwork, required an unmoved mover to flip it into action. That is, a metaphysical being called God. Was this magisterial intellect wrong? His only mistake was to believe in 'the one God' when, in fact, there are billions

69. I put on 'deacon Blue', one of my favourite bands, and sat down to type. I was feeling cold and lonely. I measured the room temperature but it was normal at 24 degrees C. I began to type but 30 minutes later I suddenly noticed that I felt very warm – almost perspiring even. I checked the room temp. and it was unchanged. I then realised that I must have formed very close PP bonds, enveloped by their rich and powerful music, with the musicians, and anyone else in the rooms surrounding my flat. I also noticed that the lonely feeling was gone. I was in a psychoplasmatic 'plenum', to borrow from the language of physics.

70. PLEASE NOTE. I thought I had spotted a contradiction in my account of the discovery of the afterlife. That is, the discovery occurred simultaneously with, and as a result of, my realization that my mother

was connected to the BPE – which is identical with the universal unconscious (UU) and she must therefore be in the afterlife. The suspected contradiction was that I would not expect to be consciously aware of my mother's unconscious which would be part of the BPE, (one is not normally aware of **one's own** unconscious) but I then realised that if I were so close I would also connect telepathically with her conscious mind, which would provide the vital clue to her identity. So all is well in that quarter, but it remains true that I, in my privileged position as a repository of this information, am still sometimes consciously aware of the BPE/UU, and it remains as a source of discomfort although I know that it will get better in time, and the universal unconscious will be truly unconscious, to me as it already is to everbody else.

71. A bond between two or more living creatures is a permanent short or long distance telepathic connection..

72. It would appear that several physicists are now strongly questioning the actuality of 'the big bang' and putting forward the idea of the 'big bounce or 'the oscillating universe' (my words) where the universe shrinks and expands. This tends to confirm my disbelief as expressed on page 44 in chapter 2 and consequently supports the 'primacy of feeling' as a fundamental principle in living systems, and the mechanics of ' judging and believing'. This is also supported by the fact that that the universe is now in accelerating mode, (bouncing away from a recent 'crunch'?).

73. It would appear that the past *is* there and one has the choice of remembering it, forgetting it or re-entering it. It seems that one's body must be a kind of permanent record or microcosm of the totality of its experience –and if one could 'get into' the appropriate mood or state, one could 'relive' one's past, as indicated in my account of a 'time-memory' experience in Coventry related elsewhere in the essay. (see addenda 42).

74. 6/01/12. In the last few weeks two articles have disappeared from my room. i.e. a pack of vegetarian leaflets, and a lid from one of my saucepans. It reminds me of the time when a chinese stainless steel knife which had disappeared, suddenly reappeared on my kitchen working surface, and the

similar occasions when a number of first class postage stamps disappeared from a hidden compartment in my wallet never to return, and a box of black pepper which vanished from my kitchen cabinet.

I was, for some time, at a loss as to where their hiding places could be, having completely ransacked my flat trying to find them, until today when it occurred to me that they must either be in some other part of the building or could, in the form of tenuous psychoplasma, be held in suspension in the CC local to myself, held ready one presumes until fortune favours their return. Note: 09/01/12. Since these words were written a strange, unrecognised and highly ornamental desert spoon has made its appearance among the cutlery in my flat. I have no idea where it came from. There is no way that it could have been put there by someone local to me. This reminds me of the time two or three years ago in my room, when I had been talking to a friend about various matters of interest. A short time after he'd left I went to a table in the centre of the room and to my surprise found a book called 'Alien Base' by Timothy Good. Its subject was UFO'S which was very surprising since I had been listening to a radio programme on this very subject that very morning or, possibly, the evening before, my memory is fuzzy here. Despite racking my brains for some time, and verifying that it was not left by my friend, (if I remember correctly) I simply could not account for its sudden appearance.

75. The GW stays virtually the same from birth to death, warm, lusty, mischievous, playful, positive; the spirit too does not change though the conscious mind, of which it is the foundation, does as a result of changes in the cognitive centre, the head. The spirit and the head together make up the conscious mind. The head and therefore the conscious mind, also change as a result of the many experiences the subject has had during his/her life. This means, effectively, that the character and personality of the subject are are liable to quite extensive change over the years. It is also possible for 'metempsychosis', the transmigration of souls, to happen and I can testify to this having, at one point in my spiritual wanderings, been occupied by not one, but three stranger spirits; I could feel them in my heart. It was a very confused and confusing time but I did, eventually, rediscover my true identity.

76. LOVE. "It feels so nice", Debby Harry with the band 'Blondie' singing of being in love with Denis.(PP auras or biological fields mix to generate feelings of being in love)

77. Yesterday, 17/01/12, a container of 'Lynx' bodyspray appeared in my flat, standing on the chest of drawers with the TV set. I haven't the faintest idea where it came from and it must join the other things listed under addenda, item 75, as inexplicable events.

78. Why does one sometimes feel down, depressed, and at other times feel inexplicably happy and elated? It depends partly on which person or people with whom one is is psychoplasmatically, and subliminally, connected at any particular moment. It could be someone next door, in the room above, and one is not consciously aware of the connection but it is there; the spirit, the self, is carried up or weighed down by subtle PP attachments. In a normal individual who has a secure central relationship and a circle of friends, this effect will be only minimally felt, but a person who does not possess these attributes may experience it as worrrying, and inexplicable, changes of mood, classifiable in some cases, as mental illness, or, naturally and positively, "I feel great today" or "I'm feeling down".

79. The spirit is responsible for all perception I think, and also exercises courage, drive, determination to attain objectives. It is the root of personality.

80. A couple of days ago, thinking about the Russian and Chinese veto's on intervention in the problems in Syria, I suddenly found myself in telepathic contact with president Putin. It was a friendly feeling and it grew stronger and stronger, so much so that I began to feel that I could influence his attitude towards Syria. Next day I heard that the Russian embargo on the Red Cross getting aid to the rebels had been lifted. It may easily have been a coincidence but it may also not.

81. I remember reading an account, author forgotten, of how the pyramids were built. It would appear that the builders used the power of

sound, I don't remember how, but I do remember that they were alleged to have tapped the blocks of stone with wands thus causing them to float into place. I mention this since it relates in a very interesting way with my computation in chapter 4 items 4 and 5.

82. I was in my kitchen. There were dozens of flies tiny, baby flies. They had been breeding in my rubbish bin and now were zooming about, climbing, swooping and diving. I was suddenly aware that I was in telepathic contact with them, part of the group. I felt a sense of pure, innocent joy as if I were a newborn baby playing and enjoying discovering my powers. I see now that this was because, in order to be connected with the BPE I had to be connected with at least one other living organism, and in this instance my link with the flies was stronger than that with any other organism since, for a few minutes, they were my most immediate and closest companions.

83. A person in the afterlife has a conscious mind, animated by the spirit, and an unconscious one animated by the god within just as in the present life. Also, as in the PL, the GW is responsible for supplying feelings to the spirit and organism in general e.g. happiness, love, fear, dislike, hope, pleasure, satisfaction, elation, relief, worry, friendliness and so on. The only difference, I think, between the AL and the PL is that the body is absent in the former. This must mean, I think, that the laws of physics do not apply: one is living in a metaphysical world.

84. As a result of recent experiences, including my memory of the kitchen of a flat in which I lived, fifty years ago in Coventry, I have concluded that memories, although stored sometimes in the body, and sometimes in the space surrounding it, must be classified in some way by the spirit or the god with in. I originally concluded (in chap. 3 item 2) that older memories were stored lower in the body i.e. the abdomen, but my experience of recalling the memory of my Coventry flat happened totally in my head or, at least, at head level. I felt that my awareness was in my head and I had a growing sense of entering a different, long departed place, which was growing more real with every second.

It would seem reasonable, therefore, that memories stored in the body as a whole relate to specific events, facts and items of information, whereas when entering the space within the head, one is entering a region which contains in material form, every single speck or item of experience to which the body has been subject. I suspect also that this fact relates strongly to the fact (chap. 6.item 4.) that time travel through PP is possible.

85. There are basically two forms of energy:

(a) physical
(b) metaphysical

Of these, metaphysical energy can be divided into:

(c) psychic energy (represented by the spirit in tandem with the head = mental energy)
(d) divine or sexual energy (represented by the god within).

86. The physical body is absent in the afterlife therefore the departed must be free of physical ailments, though they may be remembered and felt.(secondary perception). This is certainly true of the metaphysical components of the subject, so it is quite likely that mental illness can carry over into the AL. In this respect one can quote the image/feeling that I got of T.S. Eliot raising his weary head. He seemed far from happy.

87. Ultimately the reality of *any* phenomenon can be judged, only when it has been perceived and this cannot happen until it is perceived by the spirit which, in its role as the person, is uniquely responsible for all perception.

88. The TMM which stresses the metaphysical basis of psi could conceivably be overtaken in twenty years by an enlightened physics. This observation is prompted by a recent physics publication called: 'Quantum non-locality and Relativity: metaphysical intimations of modern Physics. By Tim Maudlin'. If the TMM were ignored and research continued it

would discover an entity having the same attributes as the god within, and another possessing those of the spirit. These might be given the titles Agency (a) and Agency (b), but their functions would remain identical with those of the god within and the spirit.

89. There is an old puzzle relating to identity which TMM clears up nicely. It concerns the man who has gradually lost every original bit of his body, after a lifetime of medical operations. Who is he now? Now that he is made up of hundreds of strangers? He is the spirit in the heart and he is supported by the god within, both of them metaphysical and unchanging. This personage is essentially R.Buckminster Fuller's 'phantom captain'.

90. 02/7/12. I have just now seen clearly for the first time, exactly what happened at the moment of revelation, when I became aware of the material nature of the mind. Prior to that moment, I had for years been making careful observations, taking notes, making diagrams and thinking hard, all on a conscious level of course. When the moment of revelation came one evening in Spring 2005, it came as an uprush of a clear semi-gaseous material from the centre of my chest, and I recognised it instantly as the thing that I'd been looking for for so long – **real** mindstuff – the mind itself, but even more important - the revelation was given to me by the god within in the unconscious mind. While I had been consciously seeking, the god within had also been busy, and the final revelation was not me shouting 'eureka' I've found it, but a gift from the god within which had been busy loosening knots and tying up loose ends in my unconscious mind. Not an intellectual discovery but more a damascene revelation.

91. The phenomenon of 'home advantage' in football, cricket, tennis etc. is the proven fact that a team or athlete's chances of winning are significantly increased if he is playing on 'home ground'. The phenomenon is caused, I believe, by the mingling of the psychoplasma of the 'home team crowd' with that of their heroes on the field, and is similar in some respects to the mechanism of falling in love .

92. An example of the Biological Field in action. One is on one's way to visit one's old friend Jim. . On the way one meets an acquaintance who seems very eager to talk, but one is very keen to see Jim.. In a moment of weakness -not wishing to seem impolite - one stops to exchange a few words. As he talks one tries to keep one's mind on one's objective – Jim, knowing that the 'buttonholer' is going to suggest having a drink in the nearby pub, but his words are getting a stronger and stronger hold so that one is beginning to forget about Jim.. At one point one is tempted to say goodbye and walk away without apology, but one feels that this would be impolite, so it's now or never – rudely break away or succumb to his spell. There are two forces acting here.

(a) social observancies, good manners (conscience)
(b) the biological field acting through BP combined with PP (unconscious and the GW)

The average person, of course, would be aware only of the acquaintance, his words and his own behaviour i.e. with or against the social requirement to 'be a nice person', and does not realise that a whole sea of psychoplasma now has him in its grip, and is binding him to his interlocutor.

93. POF can be seen as a final exposée of the positivist approach, the reliance on accepted science and the dismissing of anything which can not be observed or measured.

94. The principle of to Feel is to Know (TFITK) finds another instance in the joy and satisfaction felt by an artist in the creation of a great work of art. He knows it is good because it *feels* so very good. A mediocre work leaves the artist with a feeling of uncertainty, 'is it any good?'and he proffers it to the public 'tongue in cheek'.

95. The laws of physics do not apply to the god within which transcends space and time. This means, for instance, in theory, that an object existing at 9.00 am. on 12 July 2012 in birmingham can, with its help, disappear, only to instantly reappear at some other, earlier time in

Manchester on 15 July 1997. The 'flip' could be from any given point in space and time, forwards or backwards, to any other point (except the future - for reasons given).

96. It occurred to me today that one can often mentally *feel* whether a given number is a prime or not e.g. 3, 5, 7, 11, 13, 17, 19, 23, 29, 31, 37, 41, 43, 53, 59, 61 etc. For the average person this could be simply because he doesn't remember them, from his schoolboy times tables, as familiar, but one can sometimes sense that numbers such as 127 and 223 are primes even though they come outside the times tables. 251, 613 are also primes but the bigger the number the more uncertain one becomes. I think this faculty is essentially the same as that described in Chapter 2 item 12, 'the mechanism of judgement and belief', which could also explain and apply to the astonishing powers of the 'savant' who can recite the value of pi to thousands of decimal places. I have looked briefly into this condition which is sometimes related to autism and is considered to be a neurophysiological disorder and apparently its cause remains elusive, which lends more weight to my suggestion.

97. I have noticed that sometimes, when I enter my room to continue work on this theory, or some other project, although it feels uncomfortably chilly at first, an hour or so later it feels fine. It actually happened just now and I now feel pleasantly warm. A few minutes ago I measured the temp. and it was 24 degrees C. which is interesting because in the care home where I used to live, where I had few friends, was very unhappy and felt cold most of the time even when the heating was on, the ambient temp. there too, was 24 degrees C. (See addenda item 69). I also remember a worker there, a very unhappy man who was always complaining that he felt cold. This strongly supports my thinking on the functions of psychoplasma.

98. I have just now, (12/10/2012), made a startling discovery. While reading and correcting the proof of TMM I came to an entry 'biological field', under 'author's experiences of psi', of which I have no recollection and when I checked there were 14 other entries of this term, which replaced an earlier, discarded, term. The term was 'biological field' and

though I must confess that it is possible that I replaced all 14 out-dated items and have since forgotten the fact, it seems unlikely. The only alternative to this, the true one I suspect, is that I am being helped by the god within, which is getting more and more enthusiastic the closer I get to indexing and finishing. This is similar to 'winning streaks' in gambling described elsewhere in the essay and J. B. Rhine's U function, and the rediscovery of the chinese knife which I'd lost. This supposition is supported by the fact that when I came to turn on my PC twenty minutes ago, the lead had already been plugged in ready to go. I am quite certain that when I vacated and locked my room earlier, the plug was lying on the table surface. Something - the god within - is helping me again.

99. If aliens have visited Earth as recently stated by a certain important person in touch with UK officials in charge of military intelligence, (BBCbreakfast@bbc.co.uk. 5 July 10), and if they really could read his mind and knew all the secrets of our intelligence, they must certainly have known the theory of material mind and developed a highly sophisticated technology based on it. They would also be able to travel much faster than the speed of light, either by physical means, the hypothetical 'wormhole', or some technology made possible by the space and time transcending powers of the god within, in a 'non-local' universe. (See chap. 5 item 13).

100. It occurrs to me that my conception of the mind/body relationship is diametrically opposite to the one which recommends 'mens sana in corpore sano' which means 'a sound mind in a healthy body'; the opposite being 'a healthy body in a sound mind'. This is a helpful and convenient way of visualising the material mind.

101. It is interesting to consider the case of Jesus who, without doubt, had enormous parapsychological powers; i.e. the tearing of the temple curtain was almost certainly a result of the conflict between Jesus's faith and that of his friends and enemies in the multitude. The feeding of the 5,000 is also probably true (ref. 20) as was the turning of water into wine, walking on water, the raising of Lazarus, and the casting out of demons. (curing

mental illness), something which I myself in my own modest way, have done twice, and have many times felt myself capable of doing as a result of my own powers which accrue from my own knowledge and seem to be growing. (I had a distinct, though transitory, experience of second sight a few days ago).

102. Unlike the theory of relativity, which TMM partially subverts, TMM is not an exact science capable of a mathematical treatment but is simply a mass of information used to paint a self-consistent picture. The student thereof should then explore the presented material in order to get the general feeling of it, and by doing so should become aware of its truth and integrity.

103. A new possibility in retro-time travel has occurred to me. It is possible to travel into one's own past but this time travel has the disadvantage of having an automatic limit, namely the point at which one's earliest memory occurs. However, when one considers the fact of my contacts with the ancient greek philosophers and the 7,000,000 years old hominid one begins to see that time in the afterlife really is utterly different from here in the current life. We know already that the god within transcends space and time, so it follows that if one can, disencumbered of the body, travel swiftly from A to B on Earth's surface, it must be possible to travel from A in time backwards to well-known dates in the past, *but* one would have no body, only the god within and the observing spirit, so one could observe ancient Rome in full flow or look at dinosaurs. By the same reasoning, one would think, one could travel into the future, but again, one must note here that the past has all happened and is set in stone, (neverthess, I believe, in this scenario one would be able to watch it happening – my conclusion is that one could watch the battle of Hastings being fought!) while the future has not yet happpened, (unless one has absolute faith in the block theory of time), and could be seen as a tree with branches and twigs or simply an unresolved chaos. Precognition does happen though, so the future in such cases must be determined though I hold that freewill, is fundamental, which means that precognitive events must form omly a very tiny proportion of the future chaos of possibilities.

There is, however, a way round these facts which preserves freewill and simultaneously a degree of determinism. If one thinks for a moment it becomes clear that a part of the future ***must*** be predetermined since it grows out of the present which is real and solid and contains cyclical phenomena such as sunrise and sunset. (to be continued in next book).

104. Generally speaking human actions are motivated by feeling rather than rational thought, as observed by David Hume.

105. It has long been a mystery as to the mechanism which causes a distant person to turn round when looked at from behind. The Theory of Material mind explains it nicely, i.e. each person has two minds (a) the unconscious one and (b) the conscious one. (See diag. 5) The physical dimensions of both vary with time and circumstances. However, when one looks at the distant person one is deploying one's conscious mind which may be ten thousand square metres in area (10^2 x 10^2) or more, which means that its presence may be felt directly or via the collective conscious by the observed person's conscious mind hundreds of metres away. One's unconscious would be already connected with that of the observed person and others nearby via the collective unconscious. I tried to verify this explanation today (07/07/2013), by taking a pair of binoculars (mag. 7 x 50) onto the street to observe. Despite the power of the binoculars I found that very distant people, (2 to 3 hundred metres) were unmoved, even though I could see details of their dress. This suggests that contact between the material minds was not made. It is possible though, that on this busy street, the ambient PP level was swamped by the presence of other road users, which does not invalidate my explanation which I still believe to be true and must confirm in better circumstances.

106. There is an old saying, 'the wisdom of the crowd' which means that a decision leading to a particular course of action followed by the crowd is sometimes better than that made by a single individual. This fact, I believe, was first noticed by science in the late 19th century and a perfect example of it was at a side-show in a village fair where participants were invited to guess the weight of a bullock in a pen. It turned out that

the average of the summed guesses was almost always nearer the correct weight than any individidual's guess. The operative mechanism here is that the rudimentary psychic or paranormal powers possessed by all and any given person would tend to make any guess closer to the target than expected by chance, which would make the guesses 'bunch around' the true weight, so that the average would be very close to it. The active agent here would be, of course, the god within.

A related phenomenon is to be seen in the activities of the leaf-cutting ant. Each ant cuts and carries its portion of leaf independently and entirely without the guidance of an overseer, but each portion ends up in the right place. The operative agents here, again, I believe, are the gods within which tend to mutually attract to rejoin the local unconscious centred at the destination of the activity – the ant-hill - or wherever these creatures live. Each ant follows the actions of the group, with no conscious or purposive aim, and is joined to its neighbour by the god within and BP in the local unconscious. Similar reasoning applies to shoals of fish, swarms of bees and the remarkable changing patterns and formations of birds congregating before their journey to southern climes i.e.unconscious telepathy..

107. The fact that it is not possible to feel without being aware that one is feeling (TFITK) is an inescapable truth in TMM, a kind of absolute, just as, perhaps even more than, the speed of light 'C' is an absolute in the theory of relativity. It is as fundamental as is the Heisenberg uncertainty principle. This is self evident.

108. Feeling is more important and fundamental than any other human faculty.

109. It has occurred to me that although it seems likely that BP is particulate (see ref.7) it is far from certain that pure spirit is so. It would seem much more likely that spirit is a mysterious, continuous force or awareness, uniting the bioplasmatic particles if there are any. It is the person him/herself. It would also seem likely that BP is not actually the god within, but simply its medium, and the god itself is of the same nature as the spirit, insofar as it is a metaphysical, non-physical presence.

110. It has just now occurred to me that my assumption that every organism in the afterlife is there right now, all assembled as a plurality in the passing moment, is quite possibly erroneous. The god within transcends time and space so when my long departed parents seem far away, which they do most of the time these days, they may be far away in space in my home town, or far away in time having 'regressed' to a happier time as suggested elsewhere in the essay as being possible. This brings up my contact with the hominid described in 'paranormal experiences of the author' (Chap.8 item 6). When I first made contact, as written, I thought I was communicating with something deep in the past, but quickly changed my mind and realised, or so I thought, that 'she' was in the afterlife right now with all other life that had passed on. I am now beginning to see that she may well have been in the past, the god within would have made this possible as with my parents, although the contact was so strong and sudden as to suggest that she was fairly close to me timewise.

111 I heard another quotation from Einstein yesterday. When asked 'what form will the theory of everything take when it is found, he replied, "I don't know but I believe it will be simple and beautiful". I trust that TMM fulfils these requirements, even though it asserts that a theory of everything can never be found.

112. Although it is not essential to TMM, I would like to show that TMM is similar to Darwin's theory insofar as it does not contain a rigorous mathematical proof as do relativity and quantum theory, but merely supplies a picture of original, innovative discoveries and conceptions being pulled together by simple, easily understood arguments.The difficult parts in TMM are those that require the student or enquirer to almost totally abandon the world-view prevailing since Newton's day, in favour of a brand new look at the state of living things and their interactions with others, and the world at large. Both TMM and evolution convince by the weight of evidence being accepted and organised around a single main truth i.e:

(a) The Theory of Evolution by Natural selection. (Darwin) or:
(b) The god within and the spirit. (TMM)

In the case of Darwin's theory, however, it would appear that there are two rival theories of evolution. The one based on natural selection and the other, less popular, based on intelligent design.. However, if we now examine these two theories it will be seen that natural selection cannot possibly work unaided – it **needs** ID. There are two forces at work here, and *neither* can work alone and unaided. The two are complementary and need each other to work. Intelligent design supplies the spirit, the motivating force, struggle and intelligence inherent in living, and natural selection, a simple passive mechanism, supplies the automatic guiding principle which results in design.

Expressed in the terminology of TMM we can say that in any higher organism there are two forces operating in the struggle for survival: (a) natural selection and: (b) the spirit in the heart

The first one is quite mindless and automatic and caused George Bernard Shaw to complain that his spirit sank and his heart turned to a heap of sand within him (or something like that). I am currently reading Richard dawkins' 'The God Delusion' in which he inveighs heavily against 'Intelligent design', which I believe is a strong, or- more to the point - *essential* factor in the struggle for survival, because a struggle it is, and the organism is intelligent *feeling* and in the higher life forms, conscious and knowing full well that it is struggling whatever its status on the tree of life. Of course the actual mechanism that modifies its form and capabilities is natural selection but deprived of the will to survive, wherein is found intelligence, selection could not happen, and nothing could grow or change. Wanting, needing, striving and effort are essential to survival and represent 'intelligent design' in the theory of that name. The intelligence is the spirit or mind, and natural selection is the origin of design.

113. The last entry in the addenda, number 112, actually lends weight to TMM insofar as the presence of the spirit in the heart, which is a fundamental component of it, helps demonstrate the way in which intelligent design could, and in my opinion does, operate, i.e.for intelligent design to operate a creator god or supreme being is not actually necessary, although many are put off ID by the belief that such a being is.

114. It has just now occurred to me, that palmistry, along with numerology, gematria, and the whole gamut of divinatory practices and beliefs, **must** embody a certain degree of truth insofar as they can be subject to the mechanism used in addenda 1 to prove the partial truth in astrology; as long as, that is, the basic material, planets, tarot cards, tealeaves, entrails or whatever, are variables and can give various results. God's existence could not be proved or disproved this way because he is an invariable. A single entity.

115. IMPORTANT TRUTH saved from main text. If the mind were based solely in matter as current neuroscience assumes *it would not be possible to conceive a perfect sphere or straight line for matter is composed of knobbly subatomic particles and fields.* The mind is based in spirit not matter. **This amounts effectively to a proof of the existence of the spirit.**

116. Taken to its logical extension, Chap. 5 no.13 implies that space travel will one day advance beyond today's crude and wasteful mode of propulsion to instantaneous travel in a non-local universe, from point to point, over vast, even astronomical distances, to establish human outposts and then settlements on the exoplanets nearer Earth. This will happen via the metaphysical component in The Theory of Material Mind.

117. The sensitive god within resides in the stomach, the seat of feeling, and is guided by the spirit and the intellect, and the actions of the subject are guided by the favourable alignment of these three.

118. Evidence for the presence and nature of the god within in the unconscious mind. I was talking to a potential new friend in the pub but was feeling very uncertain. Could I trust him? I suddenly became aware of an amusing image, near ground level, of his penis and mine, 'entwining' as though shaking hands. I took this as a confirmation by the god within, that he was, indeed, a friend and could be trusted. I was, at the time, very lonely and isolated.

119. The god within is simply the correct title for what evey body knows as the 'something inside that will not be denied' to quote the song, or, to quote another one, 'something inside so strong', or to quote the rock band the Who, 'faith in something bigger, faith in something bigger, faith in something big inside ourselves'. It is not simply the gut itself, it is a divinity, a living presence.

120. Next item. An explanation of the function of the 'worldline' which relates space, time and motion. [Extract from INTRODUCING TIME by Craig Callender and Ralph Edney. Icon Books Ltd.ISBN 1-84046-592-1.

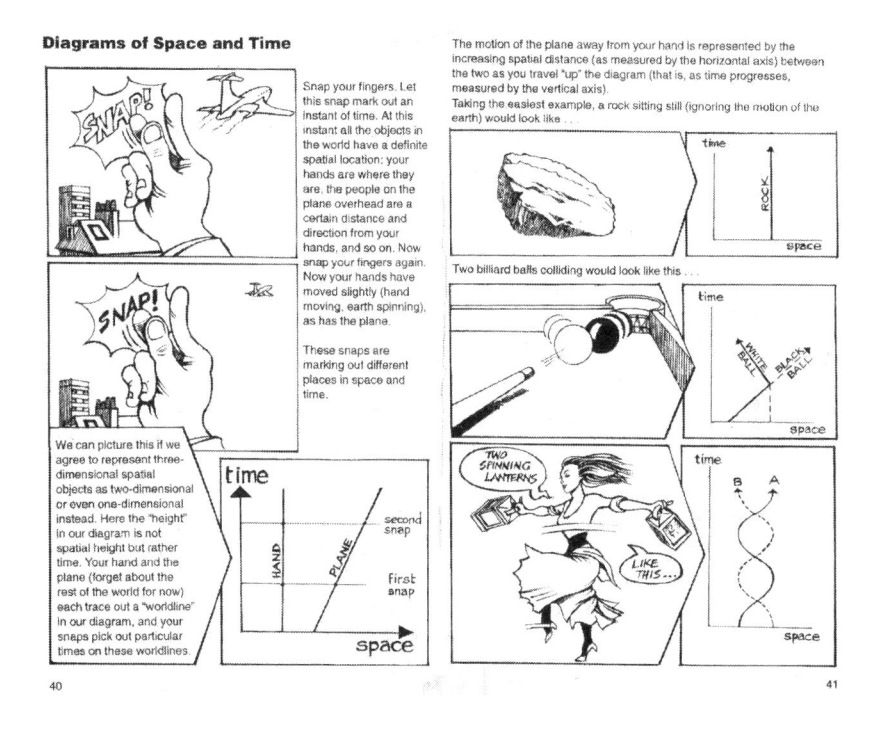

Diagrams of Space and Time

Snap your fingers. Let this snap mark out an instant of time. At this instant all the objects in the world have a definite spatial location; your hands are where they are, the people on the plane overhead are a certain distance and direction from your hands, and so on. Now snap your fingers again. Now your hands have moved slightly (hand moving, earth spinning), as has the plane.

These snaps are marking out different places in space and time.

We can picture this if we agree to represent three-dimensional spatial objects as two-dimensional or even one-dimensional instead. Here the "height" in our diagram is not spatial height but rather time. Your hand and the plane (forget about the rest of the world for now) each trace out a "worldline" in our diagram, and your snaps pick out particular times on these worldlines.

The motion of the plane away from your hand is represented by the increasing spatial distance (as measured by the horizontal axis) between the two as you travel "up" the diagram (that is, as time progresses, measured by the vertical axis).

Taking the easiest example, a rock sitting still (ignoring the motion of the earth) would look like . . .

Two billiard balls colliding would look like this . . .

121. It has just now now occurred to me that idea of using the DNA of extinct animals, to recreate them in todays's environment as in Stephen spielberg's 'Jurassic Park', would not work because it would, as explained, need the metaphysical components of the spirit and the God within, which would be absent. All one would have is a test tube of DNA – a lifeless chemical; although, thinking further, one could inject it into a

simple living cell as Venter did, and get the whole animal, although the DNA might be damaged so this could be somewhat inhumane.

122. The god within is 'Jokerman' in Bob Dylan's song and 'Sysiphus' in the ancient myth of that name.

123. I may be wrong but I believe that Plato thought that the spirit existed before the body and simply awaited birth in some suitable body. There is no doubt that the spirit does exist but I would postulate that it is created – born – at the same time as the body.

124. A baby has no habits. An old man is all habits – just a thought!

125. More on the elixir of life. I was towelling myself dry after a bath this morning when a long-forgotten name flew into my mind. I recognised it at once and in doing so was reminded of who I am – my identity. Soon after though, an old, again long-forgotten, feeling arrived in my mind and it felt clean, clear, sunny as I used to feel at age twenty. I think that this is more evidence and instruction on how to live one's life in order to stay, and look, younger. One receives and recognises in one's mind, a complex of words and feelings which carry, stronger, or weaker, the sense of 'I', that is, oneself. At such times one should be engaged in a pleasant, positive activity perhaps a hobby or an exercise and soon one will be a warm, solid reconstruction or amalgamation of all the good thoughts and feelings that one has experienced in the past. One remembers, feels, one's better, happier self, located at the midriff.

126. It is significant that attempts to communicate with the AL are basically telepathic in nature, and are subject to the same limits as experiments in telepathy here on earth, which is why successful and honourable mediums are in such short supply.

127. An imaginative picture of the afterlife and present life. Imagine a ten metre glass or perspex cube whose walls are ten centimetres thick. It is full of clear PP in which float, at various levels, and in various places, a number of entities, some the size of grapes, some peas and some melons.

Some of them lie on the bottom. Each one of them represents the mind of a person or organism and all of them have diffuse boundaries so that they merge when close together and are, in any case, connected as are spirits in the universal consciousness by the background PP or the BPE. Some of them have a material body at their centre - these are still alive in the current life – the others are in the afterlife. Telepathy can occur between the ones with hard centres but if it happens between hard and soft centres is seen as communication between AL and PL. They all freely intermix.

128. Addition to Elixir of life (chap. 5.item 7).When one is experiencing extended youth as described by TMM one is remembering various stages, states of being, of one's past life and recombining them, words, feelings and general impressions and one does not simply remember at head level one *feels* in the heart and all over and through the body.

129. It would appear that, according to the reasoning in the the TMM *everything* is real. This is not so however, it seems that the future is unreal and so is an erroneous conclusion to a process of reasoning.

130. Thinking about the big bang theory today, and the universe, which apparently, is still accelerating, a fact which I could not explain or understand, it occurred to me that the oscillating universe which is what I call it, is simply behaving as does a weight hanging on a spring and it is now accelerating, charged with kinetic energy, towards its zero-point where the pendulum is moving fastest; after that point is reached it will gradually slow down and expand to its most inflated state. After this it will start moving and will accelerate to its zero-point again after which it will start slowing to its most inflated state again, but at all times the mystery of the gods will prevail.

131. It is interesting that the images of Hume, Einstein, etc., that I saw when contacting the AL, were wearing different expressions from the pictures of them that I possess.

132. Concerning Chap. 5 item 13 it occurrs to me that the fact that the mathematics didn't reveal the truth is a reliable confirmation of my law

of the Primacy of Feeling; i.e. the logical argument was disqualified and contradicted by the experimental results so feeling wins and the law is not violated.

133. The theory and technology of TMM are destined to expand to undreamed of proportions in the next few centuries, but the theory itself provides a rock solid foundation on which all future progress can rely.

134. A few years ago I received a verbal telepathic message from a friend in the afterlife who had died of a heroin overdose. His name was Daniel and he said " it's OK up here Phil". I was, of course, delighted.

* * *

SUMMARY ANALYSIS

of already existing and newly discovered or postulated ideas and entities..

(5) Principles: The primacy of feeling. To feel is to know. Anthropic cosmological principle. The triple nature of Reality. The individuality of the spirit.
(6) Faculties: Feeling. Intuition. Instinct. Memory. Cognition. The dual nature of imagination.
(7) The organism: Head. Heart. Gut. Function of the sneeze. Function of the cough. Clearing the throat. Significance of the burp.
(8) Phenomena : Telepathy. Psychokinesis. Poltergeist phenomena. Precognition. Remote viewing. Psychometry. Dowsing. Divining the future. Teleportation. Time travel. Timeslips. The phantom limb. Making a distant person turn round. Levitation. OBEs and consciousness. Kirlian photo of missing leaf-lobe. Psychoplasmatic bonding, human-animal--vegetal.
Group minds. Psychoplasmatic blobs and streamers, funnels and tubes.

* * *

METAPHYSICS

Examples of traditional metaphysics are: Platonism, Aristotelianism, Thomism, Cartesian dualism, idealism, Kant's noumenon and phenomenon, realism and materialism.

TMM metaphysics essentially asserts the reality of the soul or spirit and the god within, and the consequent reality of the afterlife. It also asserts that reality has three aspects, and imagination two aspects. Also, of course, it gives an explanation of the mechanisms of paranormal phenomena. Its purpose is scientific not philosphical and it presents a description of things as they are, not as they seem to be or could be.

The following statements are true:

1. Consciousness exists. (Self-evident)
2. The spirit exists. I can prove this by (a) reference to item 5 later in this section, and: (b) the fact that if the mind were based solely in matter as current neuroscience assumes it would not be possible to conceive a perfect sphere or geometrical figure, therefore one must accept that mind is based in spirit, not matter.
3. The afterlife exists. Incapable of proof as yet, except by personal contact when TFITK will operate.
4. The god within exists. Incapable of proof as yet. As above.
5. In OBEs a definite 'something', carrying the sense of 'I', floats clear of the body and can pass through walls and ceilings. This has been clearly shown and proven.(Refs. 9a, 10a, 10b and 14). It is OBVIOUS, to anyone with half an eye, that this conscious entity is what has been referred to from time immemorial, as the immortal soul or spirit.

* * *

EPILOGUE

THE FUTURE OF PARAPSYCHOLOGY

I see an important and beneficent future for parapsychology, in medicine, transport, communication, and brand-new psi-based technologies an example of which is my invention 'Predictor', which will make it possible to locate lost items and missing people and even loosely predict the future. The main feature, I think, will be that a vast amount of freedom and power will be bestowed on humankind. Almost anything will become possible. Established physics-based medicine will be massively augmented by psi-based medicine which will enable precise subatomic intervention in malfunctions such as cancer, multiple sclerosis, neurone-motor disease, asthma, alzheimer's and Parkinson's disease.

It occurred to me that universities around the world could be equipped to research psi, based on the TMM, but I had forgotten that the GW would almost certainly object and would refuse to co-operate. This was depressing at first but then I saw that if, instead of a college, one built a hospital with similar research facilities, 'teleportation station' with trained operatives, a 'resurrection centre' or 'time travel departure point', that is, places which would make it possible to put psi to some practical use, it would be only too happy to oblige by supplying the necessary energy, just as it has done so far during the whole history of human endeavour, that is physical/chemical energy for actuating muscles (supplied by the physical gut), and metaphysical/emotional energy (supplied by the god within, in the gut) to fuel mentation and the feelings in general.

I also saw that the GW would certainly not object to research on the application of psi to medicine, or in fact, to anything which would lead to a decrease in suffering on Earth and an increase in the wellbeing of humans, animals, plants and life in general.

I think the New Age, for that is what the next two millennia represent according to astrology, is going to bring what can only seem like quite miraculous advances in technology and the things we can do. I was at my PC yesterday, planning a journey from Lee to Neasden by means of the internet, and trying to find a route with no changes. It occurred to me that this lugging around of one's physical body would be unnecessary in one or two centuries – one would simply call on the services of one's personal 'teleportation centre', a local firm which operated a teleportation service, and this would allow one to teleport to one's destination in a matter of minutes or even seconds, from a station or even one's home. One would imagine that in this process the CC would be flooded at any given moment, with human, animal and vegetal bodies as well as inanimate ones, all in a state of temporary suspension in the form of psychoplasma, as they travelled to their destinations. This, if it were true, brings up rather horrifying visions of mid air psychoplasmatic collisions, but it is almost certain that teleportations will be instantaneous, so this would not apply. I also saw in my imagination, an Earth with a human population of only one or two billion instead of seven, and a planet with much of it's erstwhile urban and suburban areas restored to their original wild and natural state, and populated even with flora and fauna from the various earlier geological ages, such as the Permian and Jurassic, when dinosaurs and teradactyls roamed the earth. All of these flora and fauna should be capable of resurrection within a couple of centuries I think. It can be seen that no plant or animal is truly extinct – they are all capable of resurrection and inclusion, in an Earth which could be transformed into a paradise.

It is also clear from Addendum 16 that space travel will be dramatically affected insofar as it will be possible to exceed the speed of light, by so much that it will be possible to disappear on Earth and at the same instant reappear on an exoplanet of our choice, many light years away in deep space.

* * *

BIBLIOGRAPHY (ABBREVIATED)

The Reach of the mind. J.B. Rhine. Penguin books. 1948.
Supernature. Lyall Watson. Hodder and Stoughton 1973 (by arrangement with Doubleday and co.)ISBN 0340 17368 8
Memories, Dreams and Reflections. C.G. Jung. Colins. 1967. The Fontana Library etc.ISBN 0 00 642519 4.
Introducing Quantum Theory . J. P. McEvoy & Oscar Zarate. Icon Books 2007. ISBN 978-184046850-2.
Mysticism and Logic. Bertrand Russell. Dover.2004(republication of 1918 original).ISBN 0-486-434440-0.
Introducing Consciousness. David Papineau & Howard Selina. Icon Books ISBN 1-84046-665-0.
Does God Play Dice. Ian Stewart. Penguin.1989. ISBN 0-14-012501-9.TV
ATOM (accompanying TV series). Piers Bizony forword by Jim Alkhalili. Icon books.978-1840468-73-1
Beyond Supernature. Lyall Watson. Hodder and Stoughton 1986. ISBN 0 340 38824 2.
The Holographic Universe. Michael Talbot. HarperCollins. 1996. ISBN 9 780586 091715
The Secret Life of plants. Peter Tompkins and Christopher Bird. HarperCollins 1973.ISBN 978-0-06-091587-2.
The Mind Race. Russell Targ and Keith Harary. New English Library.1986 ISBN 0-450-39008-X.
A World Without Time. The forgotten legacy of Gödel and Einstein. Palle Yourgrau. Penguin. 2005. ISBN 978-0-140-28672-4.
New Perspectives. Stuart Holroyd. Arkana. Penguin. 1989. ISBN 0-14-019195-X.

Descartes' Error. Antonio Damasio. Vintage 1995. ISBN 978-0-099-50164-0.

Einstein's Universe. Nigel Calder. Penguin 1975. ISBN 0-14-022407-6.

The 21st Century Brain. Steven Rose. Vintage 2005. ISBN 978-0-099-42977-7.

Introducing Time. Eric Callender and Ralph Edney. Icon books. ISBN 1-84046-592-1.

* * *

ABOUT THE AUTHOR

I started my working life in the electronics industry and trained in telecommunications, but switched to fine art a few years later, and am now a fully qualified practising visual artist. I studied art at Goldsmith's college school of art and was awarded an honours degree.

I have always been interested in the paranormal and my belief in telepathy was confirmed at age 17 when I had a clear telepathic communication with my mother. I was also fascinated by the symbolism of astrology - magic- despite my technological background, and my theory contains a watertight logical proof that there is a measure, at least, of truth in astrology.

In 1974 I left my job teaching art at Goldsmith's to live in my studio and deal in antiques. It was at this point that I began to get my first low-level manifestations of paranormal activity. These took the form of clicking noises emanating from the furniture which could have been due to temperature changes, but which I rapidly concluded were psychokinetic in origin. During the next 30 years or so these minor manifestations grew into extremely dramatic and varied forms, and I began to take notes and think seriously about their origin. I gradually began to see connections and a distinct pattern emerged from which I developed my theory.

The raw material of my theory, my own experiences, were utterly fantastic, and included visits from ancient greek philosophers and a seven million years old hominid, the discovery of, and visits to, the afterlife, contacts with long dead famous personages, the possibility of resurrection of the departed, teleportation, and travelling in time.

My final discovery, the root and cause of the phenomena, was the 'god within', a powerful, purposive intelligence which animates the unconscious, and whose nature and function are to supply enthusiasm and feeling to the organism. The nature of the god within is essentially good though it can be very mischievous as is shown in the book. The existence of the god within was discovered, I believe, by the ancient Greeks and their word for it was 'enthousiasme' which means, of course, enthusiasm, the god's chief characteristic. The word is derived from the ancient Greek, 'en thous', which means 'to be possessed by a god'.

INDEX

A

abdomen   1, 33, 56, 58, 103–4, 152–3, 252, 273
afterlife:
   access to gained only through dying   34, 79
   band of departed scientists could be formed to visit distant times   147
   contact made with Leonardo   84, 141, 177–8, 185
   contact with Leonardo, Beethoven, Newton, Haydn, Bertrand Russel, Koestler,
      Einstein, Jung, Hume, hominid   29, 84, 140–1, 147, 177–8, 182, 185, 217,
      256, 259, 279, 282
   contact with Leonardo, Jesus, Hitler   84, 141, 177–8, 185–6
   departed being is a psychoplasmatic presence, i.e. conscious mind plus
      unconscious (spirit + god within)   132
   departed beings float about at various levels   7
   discovery of through contact with departed mother   79, 177
   events widely separate in time can occur together   131
   is accessible through telepathy   131, 145
   is much the same as the present one except one has no body   27, 80, 148
   is resurrection made possible   131–3
   method of making contact   81
   nature of contacts same as telepathic contacts   78
   resurrection through teleport is possible   83, 222
   to slip like a tear into the shining ocean   174
   no teleport possible   148
   a thought experiment   136, 147, 150
   no verbal proof but experience of contact convinces   61, 81
age of Aquarius, the   97, 114, 147, 166, 171, 173
age of reason, the   xii, 35, 166, 171
alien spirits   22
'All gods are good but some serve evil ends'   17, 34, 58, 76, 116
'All is change except for the speed of light (physics), and the god within
      (metaphysics)'   146

'All knowledge comes from feeling'—Leonardo   41
all matter is transparent to spirit   8, 195, 198
all things, physical, mental and spiritual interact   1
Amazonian shaman   21
animals can be bored, excited, even in love   149
anosognia, disease resulting in deficiency of feeling   51
anthropic cosmological principle   83, 96, 170, 289
Aquinas, Thomas, and angels dancing on pinheads   49, 217
Aristotelian thought   166
Aspect, Alain, in Paris 1982   127
astral travelling   7, 203
astrology:
  born in ancient Babylonia, but prevalent in Renaissance, Newton was a believer   170
atheism:
  invalidated   96, 102
  unqualified is no longer tenable   96, 98
author's experience of psi phenomena:
  affects the weather   206
  'give a ghost a chance'   190
  kettle explodes   178, 181
  transmission of reel of sellotape through wall of cupboard   94, 110, 181
author's near 'timeflip' to age 14   137, 179
axiom   1
ayahuasca   21

B

Bacon, Roger   2
bacteria:
  slimes and moulds will work for us   102
basal density:
  equals 0.001 of density in heart   20, 107, 122, 125, 164
Beethoven, Ludwig van   46, 49, 130, 147, 182, 220, 255, 259, 262
Bell, John S.:
  conclusion—Nature is nonlocal   127
  inequality principle to test question raised by EPR   127
Berkeley, Bishop   51
bevy of monkeys at their typewriters   43, 45
big bang:
  is example of law of POF being violated   43
  is untrue   217

Bigley, Ken   49
biological field:
    can throw a spell over others   123
    sometimes thick and turgid   123
bioplasma and psychoplasma:
    is stuff of love and desire   10
bioplasmatic envelope:
    connects all life past and present in a single whole   217
Blair, Tony   49, 183
blobs, streamers, tubes, and funnels   154, 163–4, 289
body:
    the not reincarnated in AL   79
bonds:
    are material (PP)   241, 250, 269
    between humans, animals, and vegetation   161
    with earthworm and spider   109, 161–2
    with elder trees   190
    friendly   190
    with money plant   161
BP:
    is gravity conscious   20
    is secreted in cells of body   27, 107, 223
    is the medium of the unconscious   5, 69
BPE:
    boundary at 8km above sea level (estimate)   164
    grew slowly starting around devonian age   143, 253
    is suffused with spirit   6, 20
    joins all life on earth including the afterlife   80, 131–2
Bush, Kate   11, 28

C

Caesar, Julius   46
cannibalism   48
Carrington, Hereward:
    weighs spirit of rat (2.25 oz.) but inconclusive   74, 98
Chalmers 'hard problem,' the mechanism of consciousness   18
chemistry of romance   51
clairevoyance or second sight   71
Clauser, John at Berkeley, 1978   127
Climbie, Victoria   168
clockwork universe, the   35, 97

collective conscious, the:
  sum of all conscious minds plus a 'no man's land' 19, 37
collective unconscious, the:
  unites all life on Earth 9
communication with houseplant 109
Conibo people, Peru 21–2
conscious mind:
  can grow to cover 10,000 metre^2 of earth's surface 120
  grows to accommodate acquired characteristics 79, 85
  and GW co-operate in mechanism of judgement and belief 42
  is finite 96, 120, 123, 218
  is rooted in heart and thorax 25, 108
  and unconscious have metaphysical agency at centre 31
conscious telepathy 17, 36–7, 216
contact (telepathic) with Riz Latif, newscaster—she uttered my name! 186
Copenhagen interpretation, the 126–7
copper talisman 169, 189
creationism 98
creativity:
  relies on intuition and feeling 49, 61
crest of a wave 37

D

Darwin, Charles 2, 42, 282–3
dawn of life on Earth, the 35, 146, 243, 253
deism 97, 166, 171
density of PP:
  high in crowded pubs, cinemas, etc. 36
  shoals of fish, flocks of birds 36
  varies 7, 143
Descartes, Rene, should have said 'I feel therefore I am' 41
difficulty of replication of psi experiments, reason for 9, 158, 160, 174
digital computer, is completely predictable, animal much less so 100
discoveries and elucidations made possible by the theory 154, 158
discovery of conscious mind xi
disturbing visions of demons 21
divination, divining wand twitches by direct PK action from GW 94–5
Don Juan figure 124
dowsing and divination 69, 94
duplicate self, fallacious argument exposed 45–6
Dylan, Bob, and 'Jokerman' 24, 180, 233, 286

E

Einstein, Albert:
   believed like everyone else that Nature was 'local'   126
   Einstein/Podolsky/Rosen paradox   96, 126
   emphasized intuition   60
   spooky action at a distance   126, 128, 241
   was wary of maths   45
elixir of life:
   Chris Chataway and psychological barrier   114, 219
   if one will regain one's youth one must first understand that it is possible   113, 115
   inspiring experience of being young again   219, 286
   it is good to play CD of a young band   269, 272
   one must believe despite the evidence   219
   process can be extended to 150 and further   114
   remember feelings of youth   113
   Roger Bannister and 4 min. mile   114–15
   whole system is affected, hair grows darker, teeth grow whiter, gums grow young and strong again   113
emerging worldview unites physics and metaphysics   97
energy:
   of god within is metaphysical   19
   mental is different from physical which can be measured   76
   physical and metaphysical   274
   spiritual or psychic   17
engine of the conscious mind   2
engine of the unconscious   2, 74
Enlightenment, the   34–5, 168, 173
enthusiasm   2–3, 15, 18, 23–5, 38, 43, 58, 73
eternal youth and the elixir of life   96, 108, 219
eunuch, can be brave and aggressive   18
Everett, Kenny:
   and insulting match   12
   summoning crazy material from imagination   133
'Everything that lives is holy'—William Blake   41, 260
evolution, was teleological   166, 173–4
examples of 'TFITK'   66, 82

F

Faraday, Michael   2
faster than light travel to anywhere in spacetime   126
feeling:

classified under three headings   37
   and emotion—universal language of   48
   and instinct   41, 172
   and instinct are based in gut   41
   and intuition cannot accept big bang, though relativity's time dilation is easily accepted   44
   is conveyed by PP   64–5
   is more important and fundamental than any other faculty   62
   is real   37, 267
   physical/emotional/mental   268
   PP is real, it is a substance   65
   reveals truth where logic fails   39
   that psi event is imminent is usually true   138
   though relativity's time dilation is easily accepted   45
   will not let you down   65
fire in the belly   23, 240
Flatland by Edwin A. Abbot   46
fly's computing ability tiny fraction of a laptop but it is autonomous since it has a metaphysical component—the spirit   103
freewill:
   and the housefly   57, 103
   a measure is possessed by all and any form of life   57
Freud, Sigmund   3, 16, 75, 115, 226, 252–3
friend thinks with foot   175
funnel of PP   6

G

Gagarin, Yuri   7
Galileo   2, 137, 166, 168
Garrett, Eileen, feats performed   71–2, 151–3, 203
Geller, Uri, is not a fraud (seen by author in a telepathic connection)   151
geometrical figures are metaphysical   3
ghosts and their nature:
   accompanying fall in temp. possibly due to warm PP being sucked out of viewers by freezing cold ghost   117
   calculation of heat transferred to ghost   118
   must consist of PP at near basal density   117
god within, the:
   always warm and friendly   19–20
   animates body in tandem with spirit   100
   can be equated with the id   21

as a divinity is beyond our understanding 150
doesn't lie or make mistakes 2, 42
existing entity 23
the god within never tells a lie 2, 42, 76
has access to the unconscious of others 59
is always right 42, 160
is androgynous 20, 130, 218
is associated with appetites 2, 18–19, 43, 58
is basic vitalising agent in individual organisms 101
is constant 112, 221
is constant, warm, friendly 19
is engine of unconscious 2, 58, 74
is epitome of goodness 15
is omniscient xiii, 42–3, 53
is origin of gut-reaction 2
is purposive and knows what it is doing in the unconscious 74
is real 253
is 'something inside which cannot be denied' 23, 285
is source of friendly, loving attachments 16
is the engine of the unconscious 2, 58, 74
knowledge not limited to memories 23
lives in the gut 2, 15, 19, 22, 25, 58, 115
nature of xiii, 16, 19, 23, 59, 75, 245, 284
never makes a mistake 2, 42, 160
plays tricks on experimenter 23
in precognition registers future events if important 146, 150
primary function is to provide feeling 24
senses via BP where water/oil deposits lie 94–5
supplies energy to produce feeling 24, 293
transcends space and time 2, 15, 23, 58, 90, 150, 217, 243, 276, 278–9, 282
and true love 96
warm, enthusiastic, omnisexual, mischievous 271
will eventually provide means for interstellar space travel 43
Gödel, Kurt, his theorem, the incompleteness theorem 62
godliness 35, 80
gods, the:
    are, were, and always will be ultimate forces underlying reality 45
    eternal mystery will endure 128
Goldsmith's college 18, 156
grand unifying theory (GUT), the:
    is impossible 172

Greek philosophers, telepathic contact with   174
Grishenko, V. S.   263
group minds:
   any group mind will share a limited collective unconscious   162
   dog and family group mind   143
   group collective conscious   162

H

Harvey, William   3
heart:
   is home of spirit (essential self)   56
   is origin of courage, drive, willpower, initiative (all intentional characteristics)   57
   learning by   59
   memories are stored in it and in lungs and ribcage   59
   revealed to have neural feedback circuits associated with memory   59, 157
Heraclitus   133, 136, 193–4, 255
here and now no longer valid, there is only place and state, here and what   149
Herz, Heinrich   2
high-voltage AC electrostatic field   10
higher animals, dogs, cats, horses, all have a place in this system as does all life on the planet   71
historical whodunnits   175
Home, Daniel Dunglas, floats in and out of windows   88
hominid, 7-million-year-old skull of early human   183
Hooke, Robert   2
human about 70 watts   77
human mind is a duality (body/mind)   13
Hume, David, and metaphysics and the self   xii, 34
hunch   52
Huxley, T. H.   83, 101

I

imagination, the dual nature of   96, 108, 158, 216, 289
individuality of the spirit (IOTS)   34, 45–6, 84–5, 139, 158, 289
individuation   104
'the inner truth is greater'   34, 112, 116, 268–9
instinct:
   based in gut   41, 61, 129
   is related to feeling   61
intelligent design   97, 170, 283
interconnectedness of the universe, the   1, 97, 244

International Space Station 6, 164
intuition, operates by head, heart and gut working together 41
intuitively perceived truths 56, 62

J

Jesus, miracles are true 46
journeys to the stars will stem from TMM and god within 151
Jung, C. G. 4, 20, 75, 115, 177, 183, 216

K

Kant, Immanuel 28
Kelly, Mary 101
Kepler, Johannes 170, 265
Kirlian, Semyon Davidovitch, absent leaf lobe, experiment 10, 54
knowledge, is unassailable, old, alchemy, astrology, faded 166
Koestler, Arthur 4, 15, 18, 37, 182, 227, 252
Kubler-Ross, Elisabeth 22, 235

L

Leibniz, Gottfried 2
Leonardo:
 contact made young Leonardo feel alive 141–2
 contact with 84, 177–8, 185, 266
 two pictures, one old one young 141
LeShan, Laurence 151–3
Lethbridge, T. C. 93–4
levitation, is related to degree of 'goodness' or saintliness in the levitator 88–9
Locke, John 51
Lodge, Oliver 93
logic:
 inadvertent misuse perverts beneficent philosophy 50
 is a tool 37
logic-based human activity leads to alarm clock 172
love and psychoplasmatic bonding is result of two minds mixing 129

M

making a person turn around by staring at his back is an instance of TFITK, staring in the park 66, 124
manipulation of minds and thoughts by powerful people 124
material bond formed with houseplant 109

material mind (conscious +unconscious), can grow to cover area of one or two square kilometres or more   71
materiality of BP is origin of PK   4
materiality of mind   1, 136, 175, 220
mathematics, is logic   41, 44, 60
Maxwell, James Clerk   2
memories:
  casting the mind back   62, 133
  exist in muscles   58
  exist outside the head and body   56
  into a noncorporeal system   133
  oldest found in lower parts of body   56
  stored in body are permanent/constant while those outside the body are in a state of flux   134
mental activity takes place in head and surrounding space   32
mental and emotional activity and psychoplasma   96
metaphysical science:
  energy is basically sexual   199–200
  figures imagined at art school   106
metaphysical science
  parts of the organism go to the afterlife at death   79
Milligan, Billy:
  a multiple personality   22
mind, the:
  and matter—both exist   27
  surrounds the body   9
minimal degree of consciousness operates in vegetal life   38
Myers, Frederick, coined word 'telepathy'   69
mysterious divinities and intelligences   97
mystery, magic, and above all, feeling lie at the bottom of science, reason is ordering principle   173

N

natural selection   98, 283
nature and all living things are joined by omnipresent intelligences   35
neuron feedback circuits and memory in heart   59, 157
neutrino passes right planet   94
Newton, Isaac:
  his 'sleep'   166
  Principia Mathematica   166
  uncaused cause   28, 171

nightmare, experienced by author as a child   160
non-distance-related nature of psi communication   96, 124
nonattached divinities and spirits   98
nonreplicability of psi experiments   15, 23

O

OBE:
   spirit perceives, not brain or body   218
   subject directs herself into nearby rooms, etc.   149
objective view of Sumerians and cuneiform script, danger of becoming
       psychoplasmatically involved with them   147
objects disappear at A to reappear at B   110, 139, 182, 192, 196
occult, the   7, 147, 166, 168–70
Omar Khayyam   173
omni-love and self-love   129–30
organ of knowing or cognition is the head   41
organisational density of computer equals that of brain tissue   57
origin and nature of thought   15, 32

P

pain, is real   29
Pan   169
paranormal phenomena experienced by author   282
parapsychological technology:
   water divining, etc., author's invention, device for finding lost things, foretelling
      the future   44
past is real, future is unreal   135, 145, 150
pendulum and map (explanation)   95
Penrose, Roger   43, 205
perceptions, qualia, are metaphysical   26
person not 'in touch with his feelings'   38
phantom limb, the   15, 54–5, 289
physiological function of:
   the burp   157, 289
   clearing the throat   157, 289
   the cough   156, 289
   the sneeze   154, 289
plants have memory and can learn   101
Plato, metaphysical forms   46, 81, 99, 243
Platonism   98, 291
PP:

boundary   11
can be male or female   8, 158
density related to sexuality   24
facilitates PK, spiritual component (S) facilitates telepathy etc.   144
is exuded through skin   4, 13, 108, 121
is full of information   12, 122, 133, 249
is present in high densities around head, neck and shoulders   37
is strongly related to sexuality   11
precognition   135, 145–6, 150, 203, 222, 231, 256, 266, 279, 289
predictions and observations made by the theory confirmed to be true   157
primacy of feeling, the:
  intellect dominates at expense of feeling   39, 52
  is fundamental to all life   45
  is vital in search for truth   39
  not observed   38, 276
private personal self is found in heart   130
private universe, public universe   1, 107
psi is conducted through space by a medium, PP, basal density is sufficient for contact   125
psychokinesis, demonstrator rotates   189, 195
psychometry:
  an article of clothing must remain connected to its owner   151
  mechanisms of   149
psychoplasma:
  attachment to Beethoven   147
  contains information   40
  and friendship/loneliness   10, 217
  interacts with matter on a subatomic level   205
  we are all connected via PP   9
psychosphere, the:
  billions of cubic kilometres probably capable of accommodating infinitude of souls   134
purposive thoughts arise in the heart but thought arises anywhere in the body   58
pyramids   135–6, 150, 257, 272
Pythagoras   29, 133, 136, 193, 255

Q

quantum theory:
  the Copenhagen interpretation   126
  entanglement is a fact   128

if two systems interact they become permanently joined and any change in system A causes instantaneous change in B no matter how great the distance between   126

R

radio communication obeys inverse square law   124
ratio of BP to spirit   4–6, 37, 63
reality:
   of the divine   28, 31
   the dual nature of   15
   incorporates a Platonic worldview   31
   is ultimately metaphysical   31
   of matter   28, 105
   of perception   28
religion:
   in bad odour   111
   obstruction to progress   97
   organised   97, 168
   and Spanish inquisition   47
   use of torture approved by Pope Innocent lV in 1252   47
   William Tyndale strangled   47
Renaissance:
   and the 'Universal man'   166
replication of experiments   xiv
resurrection of body is possible   158, 258
Rhine, J. B.:
   found that interest and motivation produce results   73
   had courage to affirm existence of soul or spirit   xii, 126
Romantic movement:
   Blake, William   xiv, 41, 80, 166, 203, 256, 260
   Rousseaux, J. J.   166
Russell, Bertrand   28, 31, 41, 47, 51, 107, 182, 226, 268

S

Sartre, Jean-Paul   17
science:
   dominated society and the intelligentsia   166
   and its current apotheosis   167
   and logical-sequential perception   49
   total knowledge of Nature will never be obtained   128
scientific method, the used to develop a psi technology   146

self, the, is tripartite, definition   31, 79
sense of self and personality are based in heart and head   41
sexuality of god within   129
snekot   17
solar plexus   1, 15, 75, 95, 103, 121, 152
spirit:
  and conscious mind are variable   221
  directs energy from god within   25, 73
  directs god within to take it to a particular place   25, 73
  doesn't age, body does   112
  and god within are most real existent things   29, 31
  in the heart   xii, 2, 13, 15, 21, 23–5, 34, 58, 100, 103–4, 106, 129, 160, 174, 188, 215, 217, 253, 275, 283
  'the individuality of' (IOTS)   34, 45–6, 84–5, 139, 158, 259, 289
  information received is registered as perceptions   25
  informs the flesh   34, 54, 58
  is a presence or awareness accompanying intentionality   25
  is foundation of conscious mind   25, 157
  is fully conscious   89
  is more real than body   29
  is motor of conscious mind   2
  is origin of courage, motivation, voluntary thought   25
  is possessed by plants   101, 188
  is real and directs actions of body and mind   25
  and its rejection by science   167–8
  leaves body at death to go to afterlife (AL)   25, 159
  moody   19
  near-proof of its existence   22
  originates in heart (chest cavity)   3, 25, 38, 56
  proof of existence   31, 284
  or soul is primary though orthodox science denies its existence   89
  'spirits ride on spirits' (we get energy from each other)   76
  sticky stuff (psychoplasma)   11, 28
  and whole mind go to after life at death   99
stroboscope   198
system, lacking freewill is not alive   98

T

Tart, Charles, performed a conclusive demonstration of the objective reality of the OBE (something rises from the body)   90
in a telepathic communication there is a small flow of PP   155

telepathic contact with hominid   29, 83, 217, 279, 282
telepathy:
   close quarters   67, 123, 152, 155, 186
   is concerned with the conveyance of feelings although discrete info (words) can get through   69, 251
   it is certain that it takes place in dreams and can lead to spontaneous teleportation   71
   medium is spiritual component in PP   69
teleportation:
   first spontaneous   189
   girl appeared in bed   94, 111, 177, 182, 190
   guitar passed through several walls and ceiling   94, 110, 192, 196
   two brown-skinned boys appeared in author's room then disappeared   192, 196
tenuous mixture of BP+S = psychoplasma present since dawn of life   35, 80
tesseract, the, or the hypercube   46
Thoth, god of the moon, reckoning, writing, learning   170
thoughts are tiny feelings   40, 243
time:
   doesn't exist, change does   135
   doesn't exist in afterlife   131
   feeling of getting lost forces return   194
   travel into distant past   253
   travel into future believed impossible   256
   travel into the past (teleport through PP) is straightforward   80
   trip to Romans, pyramids, geological periods   150
   as we know it   131
timeslips   12, 93
Tipler, Frank J., book: *The Physics of Immortality*   83
TMM embodies a triplicity, body, spirit and god within   29
to feel is to know (TFITK), examples of   66, 70, 79, 82, 217

U

umbrella PK demonstrator   73, 189, 195, 198–9, 231
UMSP:
   (poltergeist phenomena) much seems to be very knowingly executed   74
   gods within can pool resources and thus move heavy furniture   235
   its mechanism   88, 158
   and kinetic energy expended   75
   ton lift goes up and down with power off   77
unconscious:
   can be caused by conflict with parental value system   73

is infinite   16, 30, 218
   is present in the gut   58, 99, 121, 215
   mind suppression phenomena   73, 158
   penetrates deep underground   94
   personal and group merge   21
   or poltergeists (UMSP)   95, 159, 246
universal conscious, shared by afterlife and present life   142
universal unconscious, the:
   and the afterlife   87, 102, 215, 270
   is identical with BPE   88, 141, 215, 270
   shared by afterlife and present life   215, 244
universe, the, is subject to the mysterious activities of the gods   172
unmoved mover, i.e. Newton's God   98, 269

V

vacuum wheel PP detector   8, 125, 197–8, 215, 245
Venter, Craig:
   and 'artificial life'   101
   his creatures will die and go to afterlife just like natural life forms   102
Voltaire   97
voluntary thought originates in heart (chest cavity), involuntary thoughts arise in
      the flesh of the body   32–3

W

Watson, Lyall   21–2, 77, 231, 264
we are interdependent entities all joined at the gut   130
we know intuitively that animals don't know that they will die   66
Welshpool   10
Western occultism has its roots in Hellenistic magic and alchemy   170
what is happening here, is happening in thousands of other places in our galaxy
      and millions in the universe   151
whatever one does, it's got to feel right   39
whole mind goes to afterlife   99
Wicca:
   English witchcraft, respect for nature, no bias against body, esoteric knowledge
      of world of spirits and unknown forces   168–9
   Robert the witch, copper talisman made by Robert   169, 189
   spells for impotence, infertility, disease, etc.   169
Wikipedia   18
Wilson, Colin   94, 120, 151, 230–1
words appear on PC before fingers touched keyboard (PK)   87

Printed in Great Britain
by Amazon